New Opportunities and Challenges in Occupational Safety and Health Management

Occupational Safety, Health, and Ergonomics: Theory and Practice

Series Editor: Danuta Koradecka
(Central Institute for Labour Protection – National Research Institute)

This series will contain monographs, references, and professional books on a compendium of knowledge in the inter-disciplinary area of environmental engineering, which covers ergonomics and safety and the protection of human health in the working environment. Its aim consists in an interdisciplinary, comprehensive and modern approach to hazards, not only those already present in the working environment, but also those related to the expected changes in new technologies and work organizations. The series aims to acquaint both researchers and practitioners with the latest research in occupational safety and ergonomics. The public, who want to improve their own or their family's safety, and the protection of heath will find it helpful, too. Thus, individual books in this series present both a scientific approach to problems and suggest practical solutions; they are offered in response to the actual needs of companies, enterprises, and institutions.

For more information about this series, please visit: https://www.crcpress.com/Occupational-Safety-Health-and-Ergonomics-Theory-and-Practice/book-series/CRCOSHETP

New Opportunities and Challenges in Occupational Safety and Health Management

Edited by
Daniel Podgórski

CRC Press
Taylor & Francis Group
Boca Raton London New York

CRC Press is an imprint of the
Taylor & Francis Group, an **informa** business

First edition published 2020
by CRC Press
6000 Broken Sound Parkway NW, Suite 300, Boca Raton, FL 33487-2742

and by CRC Press
2 Park Square, Milton Park, Abingdon, Oxon, OX14 4RN

© 2020 Taylor & Francis Group, LLC

CRC Press is an imprint of Taylor & Francis Group, LLC

Library of Congress Cataloging-in-Publication Data

Names: Podgórski, Daniel, editor.
Title: New opportunities and challenges in occupational safety and health management / edited by Daniel Podgórski.
Description: Seventh edition. I Boca Raton, FL : CRC Press, 2020. I Series: Occupational safety, health, and ergonomics : theory and practice I Includes bibliographical references and index.
Identifiers: LCCN 2020005684 (print) I LCCN 2020005685 (ebook) I ISBN 9780367469320 (hardback) I ISBN 9780367505325 (paperback) I ISBN 9781003050247 (ebook)
Subjects: LCSH: Industrial safety--Management. I Industrial hygiene--Management.
Classification: LCC T55 .N4626 2020 (print) I LCC T55 (ebook) I DDC 658.3/82--dc23
LC record available at https://lccn.loc.gov/2020005684
LC ebook record available at https://lccn.loc.gov/2020005685

ISBN: 978-0-367-46932-0 (hbk)
ISBN: 978-1-003-05024-7 (ebk)

Typeset in Times
by Deanta Global Publishing Services, Chennai, India

Contents

Editor

Daniel Podgórski graduated from the Faculty of Electronics of the Warsaw University of Technology in 1983, and received his PhD at the Institute of Industrial Automation of the Faculty of Mechatronics of the same university in 1992. Since 1984 he has worked for the Central Institute of Labour Protection (currently CIOP-PIB), initially as a researcher in the Department of Ergonomics and the Department of Safety Engineering, and then, as the deputy director for management systems and certification. The scope of research studies carried out by Dr Daniel Podgórski mainly included the analysis of factors influencing the implementation of OSH management systems in enterprises, and the application of selected key performance indicators (KPIs) for the measurement of operational performance of OSH management processes. His competences in the field of systematic OSH management are also being used in his function as chairman of the technical committee TC 276 for OSH Management Systems, which operates within the structures of the Polish Committee for Standardization (PKN). In recent years, the scope of scientific interests of Dr Daniel Podgórski has been extended to include the development and implementation of novel digital technologies for improving safety and health at work, which comprises in particular the use of smart personal protective equipment (smart PPE), workplace wearables and other Internet of Things technologies to facilitate OSH risk management at Industry 4.0 workplaces.

Contributors

Zofia Pawłowska graduated from the Faculty of Mechatronics of the Warsaw University of Technology in 1975, and in 1982 she received her PhD at this Faculty. Since 1996 she has worked at the Central Institute of Occupational Protection (currently CIOP-PIB) as a researcher in the Department of Occupational Safety and Health Management. The scope of research carried out by Dr Zofia Pawłowska mainly covers methods aimed at the improvement of OSH management systems' performance in terms of ensuring better prevention of accidents and diseases at work. In recent years, her research studies has also included exploration of links between OSH management and workforce age management, and the introduction of social responsibility principles into OSH management systems. As a member of the technical committee TC 276 for OSH Management Systems, and the chair of the technical committee TC 305 for Corporate Social Responsibility (CSR), both operating within the structures of the Polish Committee for Standardization (PKN), she actively participates in standardization activities in these fields. She also provides advice and conducts numerous training courses for employers and managers, with a focus on issues related to her research and area of competence.

Małgorzata Pęciłło graduated from the Faculty of Marketing and Management of the Warsaw School of Economics (SGH) in 1997, and obtained her PhD in 2007 at the Institute of Organization and Management in Industry (ORGMASZ, Warsaw). Since 1999 she has been employed as a researcher at the Department of Occupational Safety and Health Management of the Central Institute for Labour Protection – National Research Institute (CIOP-PIB). In her work, she focuses mainly on the development of new occupational health and safety management methods and their application in practice, involving in particular process approach to OSH management, the use of Balanced Scorecard, the development of programs for modification of workers' hazardous behaviours, and the exploitation of resilience engineering concepts. Dr Małgorzata Pęciłło is the author of many educational materials and gives lectures at post-graduate studies and seminars in the field of OSH, organised for employers, safety managers, and workers alike. For many years, she has also conducted consulting activities covering the implementation of effective risk assessment procedures and accident prevention programmes in industrial enterprises.

Anna Skład graduated from the Warsaw School of Economics (SGH) in 1996, where she studied international economic and political relations. She continued her education at SGH participating in the postgraduate studies for quality managers (2007 2008), and doctoral studies in business management (2009–2013). Between 2001 and 2006, she worked in the IT section at Orbis S.A. (tourist services), and in the years 2006–2008 she was implementing the integrated management system in the Marshal's Office of the Mazovian Region. In 2008 she started working for Orange telecommunication company in the area of designing and improving business process architecture. Since 2014, she has been employed in the Department of Occupational Safety and Health Management at the Central Institute for Labour Protection – National Research Institute, where in 2017 she defended her doctoral thesis on the implementation of fuzzy cognitive maps for the improvement of OSH management systems. Her scientific interests focus, among others, on the measurement of OSH management system effectiveness, the implementation of process approach in management systems, age management practices, and the introduction of the concept of vision zero with regard to accidents at work.

Series Editor

Professor Danuta Koradecka, PhD, D.Med.Sc. and Director of the Central Institute for Labour Protection – National Research Institute (CIOP-PIB), is a specialist in occupational health. Her research interests include the human health effects of hand-transmitted vibration; ergonomics research on the human body's response to the combined effects of vibration, noise, low temperature and static load; assessment of static and dynamic physical load; development of hygienic standards as well as development and implementation of ergonomic solutions to improve working conditions in accordance with International Labour Organisation (ILO) convention and European Union (EU) directives. She is the author of over 200 scientific publications and several books on occupational safety and health.

***

The "Occupational Safety, Health, and Ergonomics: Theory and Practice" series of monographs is focused on the challenges of the 21st century in this area of knowledge. These challenges address diverse risks in the working environment of chemical (including carcinogens, mutagens, endocrine agents), biological (bacteria, viruses), physical (noise, electromagnetic radiation) and psychophysical (stress) nature. Humans have been in contact with all these risks for thousands of years. Initially, their intensity was lower, but over time it has gradually increased, and now too often exceeds the limits of man's ability to adapt. Moreover, risks to human safety and health, so far assigned to the working environment, are now also increasingly emerging in the living environment. With the globalisation of production and merging of labour markets, the practical use of the knowledge on occupational safety, health, and ergonomics should be comparable between countries. The presented series will contribute to this process.

The Central Institute for Labour Protection – National Research Institute, conducting research in the discipline of environmental engineering, in the area of working environment and implementing its results, has summarised the achievements – including its own – in this field from 2011 to 2019. Such work would not be possible without cooperation with scientists from other Polish and foreign institutions as authors or reviewers of this series. I would like to express my gratitude to all of them for their work.

It would not be feasible to publish this series without the professionalism of the specialists from the Publishing Division, the Centre for Scientific Information and Documentation, and the International Cooperation Division of our Institute. The challenge was also the editorial compilation of the series and ensuring the efficiency of this publishing process, for which I would like to thank the entire editorial team of CRC Press – Taylor & Francis.

***

This monograph, published in 2020, has been based on the results of a research task carried out within the scope of the second to fourth stage of the Polish National Programme "Improvement of safety and working conditions" partly supported – within the scope of research and development – by the Ministry of Science and Higher Education/National Centre for Research and Development, and within the scope of state services – by the Ministry of Family, Labour and Social Policy. The Central Institute for Labour Protection – National Research Institute is the Programme's main coordinator and contractor.

1 Introduction

Daniel Podgórski, Zofia Pawłowska,
Małgorzata Pęciłło and Anna Skład

Knowledge and practice in the field of occupational safety and health (OSH), including systematic approaches to OSH management, have been developing for several decades. This topic has been the subject of many studies, scientific articles, dissertations, monographs, etc., as well as the basis for the development and dissemination of many practical guidelines and specific standards, both at national and international level. At the same time, it should be noted that the global economy and its business environment, and thus the world of work, has recently been strongly influenced by demographic and social changes and globalisation, as well as the rapid development and introduction of novel, sophisticated and previously unknown technologies and new business models, especially in the context of the so-called fourth industrial revolution (otherwise known as Industry 4.0). These changes pose a number of challenges in terms of maintaining and improving the effectiveness of OSH management, as in new working environments traditional approaches may no longer be effective.

The word "challenge" has different meanings depending on the context in which it is used, but in business and management this term is usually used in the sense of a problem that stimulates a person or group to take specific action (Merriam-Webster 2019a). Thus, challenges should not be seen as obstacles but rather as stimuli to search for new solutions in order to make improvements or a progress in a particular area. This in turn leads us to use the term "opportunity", which can be defined in this context as "a good chance for advancement or progress" (Merriam-Webster 2019b).

Furthermore, when considering the role of opportunities in management activities, including improving the effectiveness of OSH management, attention should be paid to the new requirements that were promoted internationally beginning in 2018 through the adoption and publication of the new standard ISO 45001 (ISO 2018). It is appropriate to cite the contents of Clause 6.1.2.3:

> The organization shall establish, implement and maintain a process(es) to assess:
>
> a) OH&S* opportunities to enhance OH&S performance, while taking into account planned changes to the organization, its policies, its processes or its activities and:
>   1) opportunities to adapt work, work organization and work environment to workers;
>   2) opportunities to eliminate hazards and reduce OH&S risks;
> b) other opportunities for improving the OH&S management system.

---

* The term "occupational health and safety" (OH&S), which was adopted in the standard ISO 45001:2018, has the same meaning as the terms "occupational safety and health" (OSH) and "safety and health at work". These latter terms are used interchangeably throughout this book.

The afore-mentioned circumstances imply very clearly that today's managers should exploit the most recent opportunities in the field of OSH management when taking ambitious measures to prevent work-related accidents and diseases, particularly when striving to improve the efficiency and competitiveness of their businesses as well as adapting to increasing trends in sustainable development. In particular, it is worth-while to explore and exploit these opportunities – the results of latest research – to ensure that they are compatible with the latest trends in business management and combine them with the introduction of novel digital technologies harnessed to realise the vision of smart manufacturing and the digitally connected world of work.

Moreover, managers in increasingly complex business environments have to face trade-offs as a natural consequence of existing multi-role management systems. Thus, safety managers as well as CEOs need to gather knowledge regarding solutions and tools that may easily and effectively assist them in OSH-related decision-making processes.

Therefore, this book focuses on five thematic areas which have been identified by the co-authors as relevant for consideration of these challenges and related opportunities. These areas include:

1) introduction of the process approach to OSH management
2) improving OSH management systems using the fuzzy cognitive maps method
3) introduction of the strategic thinking approaches in relation to OSH management
4) integration of OSH management within the framework of the CSR concept
5) enhancing OSH management processes through the use of smart digital technologies.

All of this is introduced here briefly, and then will be addressed individually in greater depth in subsequent chapters.

First, the inherent objective of business activity is to generate economic profit and increase the value of an enterprise. Although in the 21st century no one questions the need to prevent occupational accidents and diseases, unfortunately ensuring safety and health at work is still seen by many business leaders as an additional burden that hinders the achievement of business goals. The opportunity to meet this challenge lies in the deeper integration of OSH activities into day-to-day operational activities. This is feasible when the traditional hierarchical organisational structure of an enterprise management is complemented by the establishment and implementation of a well-thought-out and structured business process architecture that embraces OSH processes and ensures their smooth integration with operational ones. This approach is particularly relevant in the context of the implementation of ISO 45001 standard, according to which an adoption of a process-oriented management is the basic pre-requisite for OSH management system compliance with this standard. Therefore, Chapter 2 introduces readers to the basic principles of the process approach in the area of OSH management. The typology of processes, in which OSH processes and their basic components are addressed and examined in order to understand their performance and inter-relations, as well as determine the skills that are necessary for effective management of OSH processes, is presented. Attention is paid to the measurement of processes effectiveness, with a particular focus on the principles of creating and selecting a set of indicators and their implementation to support decision making in an organisation. The Activity-Based Costing (ABC) method and the possibilities resulting from its application in the economic analysis of process costs are also presented in Chapter 2.

Since the standards setting out the formal requirements for OSH management systems were established, the concepts of these systems have been the subject of lively discussion, with both supporters and opponents arguing whether the implementation and improvement of such systems contribute to the real improvement of workers' safety and health. Without taking a particular side, Chapter 3 attempts to showcase a new advanced method for improving OSH management systems in order to credibly ensure that they are more effective in preventing work-related accidents and diseases. The chapter describes the method of fuzzy cognitive maps (FCM) and the means of applying this method to improve OSH management in enterprises. The FCM concept is based on an assumption that although people possess valuable knowledge resources, these resources are not used in practice due to the lack of user-friendly methods for extraction and further processing. There are many scientific studies proving the use of this method in management sciences and confirming its effectiveness in solving practical problems. Chapter 3 describes in detail the subsequent stages of creating a model of an OSH management system in an enterprise via the use of FCM, explains the expert role of workers, and then proposes ready-made linguistic scales for the assessment of objects and influences adopted in an FCM-based management system model. The chapter also covers how to transform a model developed by experts into a mathematical record, apply this record in simulations, and use the achieved results in the decision-making process.

The search for new solutions to improve OSH management performance has led to the conceptualisation and increasingly widespread use of process approaches. However, the introduction of these approaches without linking the processes and indicators developed to the enterprise's global strategy is useless when trying to understand the enterprise management system as a whole. It is therefore necessary to implement a holistic approach which would make it possible to link the processes carried out with the enterprise's strategy as well as with respective resources applied together with its intellectual capital. Hence, in Chapter 4, linking the performance of OSH management processes with strategic management assessment is discussed in detail. The application of the strategic management performance metric called Balanced Scorecard (BSC) to OSH management as an assessment and diagnosis tool is described. This chapter also explains how the BSC may facilitate the implementation of the process approach as well as the implementation of traditional preventive programs, such as behavioural-based safety (BBS). Finally, Chapter 4 presents the issue of senior management perception of the importance of safety and health when making trade-offs. This in part is based on the resilience engineering theory, a new approach to OSH management developed to find a counterweight to overall managerial practices that typically focus on achieving effectiveness and process optimisation.

Another answer to these challenges involves a new approach to extend the traditional area of OSH management so that it not only focuses on the prevention of occupational accidents and diseases but also includes measures to improve workers' health and well-being as well as prolong their working lives. This approach requires that Corporate Social Responsibility (CSR) aspects be taken into account in the management of OSH, including the incorporation in this context of appropriate measures to reduce psychosocial risks in the working environment and to take account of the diversity of workers, particularly the ageing of the workforce. By managing OSH in a socially responsible way, organisations may contribute to the achievement of the United Nations Sustainable Development Goals, particularly with regard to the eighth goal, which relates to the promotion of inclusive and sustainable economic growth, employment and decent work for all (UN 2015). With regard to these aspects, Chapter 5 addresses OSH management challenges from a social responsibility perspective. OSH-related issues of CSR are identified in accordance with the ISO 26000 standard "Guidance on Social Responsibility" (ISO 2010), and opportunities to integrate these issues into OSH management system components are discussed, taking into account the system structure adopted in the ISO 45001 standard. Chapter 5 pays particular attention to systematic measures to maintain workers' work ability and to integrating the age management into socially responsible OSH management, particularly in light of emerging demographic changes in society. Finally, the chapter discusses the method of evaluating the outcomes of socially responsible OSH management and presents a set of performance indicators that can be used for this purpose.

The basic rationale behind the content of the final thematic area is the recently observed radical progress in the development of manufacturing technologies, in particular the increasing involvement of digital technologies in manufacturing and business processes. This movement, often referred to as the digital transformation of the economy, is based on the application of new concepts such as the industrial Internet of Things (IoT), cyber-physical systems (CPS), collaborative robots, cloud and edge computing, advanced methods of data analytics and so forth, whose introduction to the workplace inevitably changes work processes and poses new challenges to OSH. In contrast, enabling technologies, which play a significant role in the Industry 4.0 developments, can be successfully deployed to create new solutions and systems to effectively support OSH management systems in enterprises. Chapter 6 presents and discusses the potential for applying digital technologies to manage safety and health at work. In particular, this chapter provides a typology and review of smart personal protective equipment and other wearables for OSH-related applications, with a emphasis on their usefulness in the context of IT-empowered OSH risk management. These issues are complemented by a vision of the future potential of using machine-learning algorithms and big data analytics methods for OSH management. The chapter also includes a discussion on privacy protection and cyber-security aspects, both of which are an important problem and to some extent a hindrance to the dissemination and application of otherwise useful digital technologies in the workplace.

REFERENCES

ISO (International Organization for Standardization), 2010. *ISO 26000:2010, Guidance on Social Responsibility.* International Organization for Standardization, Geneva, Switzerland.

ISO (International Organization for Standardization), 2018. *ISO 45001:2018, Occupational Health and Safety Management Systems – Requirements with Guidance for Use.* International Organization for Standardization, Geneva, Switzerland.

Merriam-Webster, 2019a. Challenge. https://www.merriam-webster.com/dictionary/challenge (accessed December 23, 2019).

Merriam-Webster, 2019b. Opportunity. https://www.merriam-webster.com/dictionary/opportunity (accessed December 23, 2019).

UN (United Nations], 2015. United Nations Sustainable Development Goals. https://www.un.org/sustainabledevelopment/sustainable-development-goals/ (accessed January 17, 2020).

2 Application of Process Approach to OSH Management

Małgorzata Pęciłło and Anna Skład

CONTENTS

2.1 OSH-RELATED PROCESSES – DEFINITION AND TYPOLOGY

The process approach to management is not a new concept. Taylor (1856–1915), who was the first to draw attention to working procedures, can be considered the prophet of such an approach. However, the origins of the concept of business process management are usually found in the qualitative movements that started in the 1950s in Japan (Ishikava 1985; Deming 1986; Juran 1989). Business processes are recognised in the literature as "a significant contributor to achieving an organisation's objectives through the improvement and ongoing performance management" (Jeston and Nelis 2014), and the approach itself as "an integrated system for managing business performance by managing end-to-end" (vom Brocke and Rosemann 2015). Although it is difficult to imagine that the process approach used for many years by enterprises in its various functional areas (including mainly production or quality management) would bypass the area related to ensuring occupational safety and health (OSH), the ISO standard on occupational safety and health management systems (ISO 2018), published in 2018, set the need to introduce such an approach to OSH management.

The word "process", from the Latin *processus* is a neutral concept with a very broad meaning, used in all fields of science. The Cambridge dictionary defines the word process

7

as "a series of action that you take in order to achieve a result". In management sciences, the term process covers every sequence of activities carried out in an organisation, i.e. both activities related to the production of material goods (i.e. production processes) and those not directly related to the production of goods (i.e. organisational processes). In the latter, information and documents, and also views and attitudes, such as those of workers or customers, may be transformed. Deming emphasises that every action is part of the process and is not an independent creation. The process is divided into steps, in which a specific work is done that changes state, input turns into output (Deming 1986).

The term process is defined no differently in the ISO 45001 standard (ISO 2018). It states that this term will be understood as a set of interrelated or interacting actions that transform inputs into outputs. In relation to safety and health processes, the inputs can be the workers' attitudes towards safety, awareness of threats or the tendency of workers to engage in unsafe behaviours. The quoted definition of the occupational safety and health process indicates two essential conditions for the existence of a business process: first, the process consists of an ordered set of activities; second, a new value is created as a result of these actions. Assuming that the creation of added value is a sine qua non of the process means that a sequence of what are known as idle actions that do not transform the input data is not a process. Such an approach is the most appropriate from the point of view of the effectiveness of occupational safety and health management.

The concept of organisational process is often identified and used interchangeably with such terms as business process, economic process, management process or simply process (Kueng 2000). Sometimes the terms are given a narrower meaning. For example, the term business process or economic process may refer only to those processes that link the enterprise to its legal, social or economic environment, and management processes may only be applied to processes related to the performance of management functions (Rummler and Brache 1995). Therefore, there is no terminological regime in this area, but it is possible to classify processes according to different criteria on the basis of literature, and the classification below does not exhaust all possible divisions (Table 2.1).

The ISO 45001 standard (ISO 2018) lists among the processes of key importance for the effectiveness of an OSH management system the processes carried out for:

- consultation with and participation of workers;
- planning;
- identification of hazards and assessment of risks and opportunities;
- elimination of hazards and reduction of risks occurring in the organisation;
- identification and access to legal and other requirements;
- internal and external communication;
- introduction of changes, both temporary and permanent;
- supervision of supplies and services and outsourcing;
- preparation for and response to emergencies;
- monitoring, measurement, analysis and evaluation of results;
- assessing compliance with legal and other requirements; and
- reporting and investigating of incidents and non-conformities and taking corrective action.

TABLE 2.1

Classification of organisational processes including safety and health management processes

Criterion	Types of processes	Place of safety and health management processes in the organisation
Type of actions carried out in the process	Processes related to product development, production, human resources development, payment execution, etc.	Processes related to ensuring safe and healthy working conditions
Added value created by processes, i.e. the significance of the process in relation to the final product (Keen 1997)	• Main processes (also called strategic, key, underlying, basic or operational) creating added value in the organisation • Management and support (auxiliary) processes that do not create added value in the organisation but are necessary for its functioning	The management and support processes include, inter alia, processes related to ensuring safe and healthy working conditions, and system processes – related to the operation of, for example, occupational safety and health management systems
Degree of association with the achievement of the organisation's objectives (Twaróg 2002)	• Basic processes (directly contributing to a specific objective of the organisation) • Regulatory processes (giving and maintaining the overall direction of the organisation, e.g. processes related to management functions)	In OSH management systems the basic processes include operating processes according to ISO 45001 (ISO 2018) and an occupational risk assessment process – regulatory processes include typical system processes related to motivation or competence assurance
Degree of association of the process with the customers (Keen 1997)	• Front-office processes directly related to the customer, inter alia order execution processes • Back-office processes not directly related to the execution of customer satisfaction	Processes related to ensuring safe and healthy working conditions are part of back-office processes; however, when considering workers as internal customers, as proposed in the Balanced Scorecard concept, the processes related to ensuring safe and healthy working conditions become front-office processes
Type of customer (Melin 1999)	• Primary processes – processes or sub-processes introduced to meet the needs of specific customers • Secondary processes – processes or sub-processes resulting from the former, performed to meet the needs of potential customers	The processes can also be distinguished in a similar way according to the type of stakeholders – in this case OSH management processes should be treated as primary processes from the point of view of workers, and as secondary processes from the perspective of potential visitors to the workplace

(Continued)

TABLE 2.1 (CONTINUED)
Classification of organisational processes including safety and health management processes

Criterion	Types of processes	Place of safety and health management processes in the organisation
Degree of process complexity (Gage 1993)	Elementary processes, complex processes and sub-processes	Occupational safety and health management process, as any process, can be decomposed into sub-processes or actions, and can be combined into complex processes
Process owners and their location in the organisation (Davenport and Short 1990)	• Internal processes – the owners are organisational units of a given organisation • External processes – involving external entities • Intra-branch processes – introduced entirely by one person or one organisational unit • Cross-cutting or general organisational processes – carried out by several organisational units or processes carried out by a team established for this purpose, consisting of workers of various organisational units	A typical example of an internal process is the process of identifying training needs, an external process – an external communication process and a cross-cutting process – an occupational risk management process
Methods of manufacturing process of products and services (Harry and Schroeder 2006)	• Manufacturing processes (when at least 80% of the value of a product or service is manufactured with the use of machinery and equipment) • Service processes (if the process is at least 80% based on human actions)	Occupational safety and health management processes are service processes
Type of objects processed in the process, i.e. by the type of input in the process (Davenport 1993b; Juran 1989)	• Processes that process material objects (e.g. manufacturing processes) • Documentation and information processes	Occupational safety and health management processes are documentation and information processes
Impact of processes on the dynamic balance of the organisation (Twaróg 2002)	• There are essentially two types of processes: • those that do not permanently compromise this balance • those that permanently compromise the dynamic balance of the organisation, i.e. what are referred to as development processes	Occupational safety and health management processes are processes that do not permanently compromise the dynamic balance of the organisation

Application of the process approach to management includes actions related to identification, mapping (visualisation), introduction and optimisation (improvement) of organisational processes. The starting point for introducing a process approach to management in an organisation is therefore to describe the introduced business processes and to present their inter-relationships, usually in a graphic form, i.e. to create a process map. In each process the following elements can be distinguished: start; end; inputs (depending on the nature of the process, these can be both products and services, as well as documentation and information); outputs (similarly to inputs, these can be both products or services, as well as documentation and information); actions carried out in a given process and the participants (actors) in the process (the process leader (the owner), i.e. the person responsible for the process; the process customer (the recipient), i.e. the person who receives materials and information or service from other people inside or outside the organisation; and the supplier, i.e. the person who delivers materials and information or service to other people inside or outside the organisation). The literature also includes the concept of a process team, i.e. a team of people participating in a process, which usually includes the process leader and his/her subordinates, participating in the introduction of a given process (Kueng 2000). A typical example of a process team involved in occupational safety and health management processes is the risk assessment team. Elements of a typical organisational process on the example of the risk assessment process are presented in Figure 2.1.

Organisational processes connect with each other to form a network (chain) of processes in the organisation. In such a network, the owner of one process may at the same time be a customer of the previous process and a supplier in relation to the next process, while the output in one process may be an input in the next, subsequent one. An example of a part of the chain of occupational safety and health management processes is shown in Figure 2.2, where the risk assessment team is the customer of the risk assessment process and is also the supplier of the recommendations resulting from the assessment for the next process of introducing the recommendations developed.

Processes can be combined into a single whole, i.e. into cross-cutting processes: the processes introduced at a lower level then become sub-processes of a higher-level process. These are often carried out by different organisational units, cutting

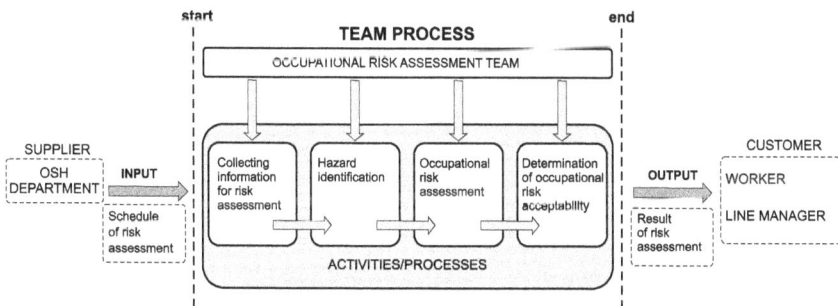

FIGURE 2.1 Elements of the organisational process

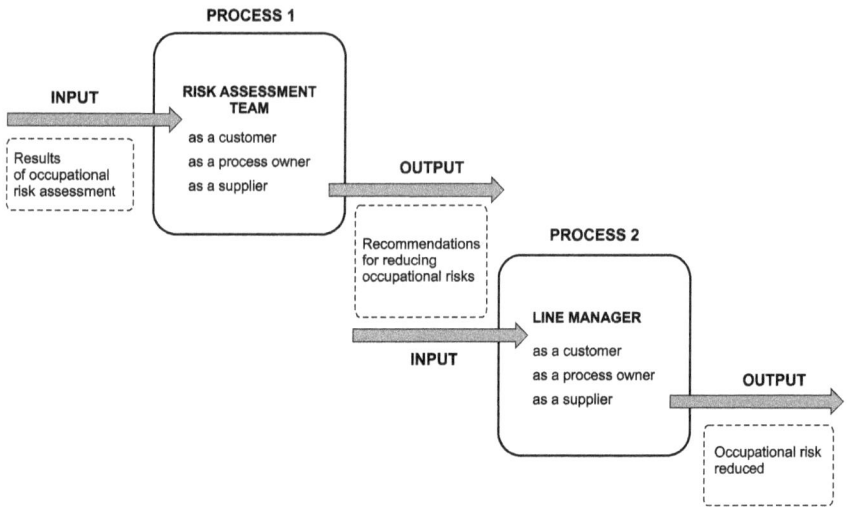

FIGURE 2.2 Examples of links between process 1 (planning of preventive actions) and process 2 (implementation of preventive actions) in an organisation

through the organisational structure. In the example below (Figure 2.3) the process of "identification of legal requirements", which has been undertaken as a result of the risk assessment, is initially carried out by the Legal and Organisational Department, then by the OSH department and finally by OSH Management System Department. At the same time the end of this process is the start of another one, "implementation of new legal requirements", also carried out by different organisational units.

In practice, if when describing processes in an organisation there is no software which facilitates process mapping, it is necessary to focus on the most important elements, i.e. the implemented actions and inputs and outputs, and the graphic form can be replaced with, for example, a table (Table 2.2).

The number of identified and described processes in an organisation and the level of their decomposition into particular actions or sub-processes depends on the level of generalisation on which the processes will be described adopted in the organisation, and on the specificity of the organisation, its complexity and the objectives of introducing the process approach in the organisation. Management theorists do not agree on the number of key processes that should be managed in an organisation, instead they indicate from several to as many as 100 such processes (Kaplan and Murdock 1991; Davenport 1993a; Keen 1997).

It should be underlined that the application of a process approach to the management of an organisation does not fully mean withdrawing from the classic functional dependencies described in the organisational structure. The organisational structure is the basis for proper allocation of resources, including allocation of expertise. However, as Rummler and Brache (1995) emphasise, as an organisation develops and becomes more complex, there are problems with co-ordination and co-operation between the different organisational units. Organisations then have to deal with what is known as the "silo effect": the objectives of the organisational units are set independently of each other, and lower-level managers start to perceive workers of other

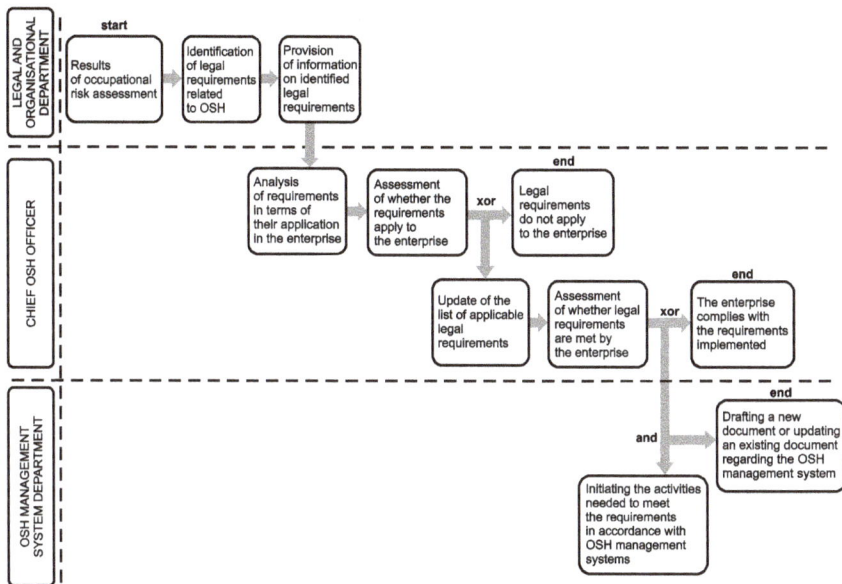

FIGURE 2.3 Identification of legal requirements – an example of a cross-cutting process

TABLE 2.2
Basic elements of the occupational safety and health risks and opportunities assessment process

Inputs	Actions	Outputs
• Information on hazards • Information on the risk protection measures applied • Information on persons exposed to risks • Information on opportunities	• Identification and documentation of methods and criteria for assessing risks associated with the identified hazards • Conducting a risk assessment • Analysis of opportunities and identification of opportunities to implement actions aimed at eliminating threats or further reducing risks • Documentation of the results of the risk assessment carried out and the resulting action plans to eliminate hazards or further reduce risks	• Documented information on methods and criteria for assessing risks associated with identified hazards • Documented information on the risks associated with the hazards • Action plans to eliminate hazards or further reduce risks

Reprinted, with permission, from: Pawłowska and Pęciłło 2018, Doskonalenie zarządzania bezpieczeństwem i higieną pracy z uwzględnieniem wymagań i wytycznych normy międzynarodowej ISO 45001. (Improving occupational safety and health management with regard to the requirements and guidelines of international standard ISO 45001). CIOP-PIB. p. 30.

units as enemies. They start building "silos", i.e. tall structures, without windows, with thick walls preventing co-operation between individual departments and solving problems together. Then there is a need to look at the organisation from a different perspective: horizontally, based on organisational processes.

2.2 MEASUREMENT AND IMPROVEMENT OF OSH-RELATED PROCESSES

2.2.1 INTRODUCTION TO THE IMPROVEMENT OF OSH-RELATED PROCESSES

The modern approach to improving organisational processes were developed in the 1990s. At that time, two basically opposite directions appeared: one initiated and propagated by Hammer (1990), the other by Davenport (Davenport 1993b; Davenport and Short 1990). Hammer advocated a radical and complete redesign of the processes in the organisation using a "from scratch" approach to achieve radical, rapid improvement. This approach was one-off in nature and did not provide for consultation of workers before the changes. Davenport, on the other hand, presented a more evolutionary approach to process improvement, shaped in the Total Quality Management (TQM) and consisting of introducing small changes while maintaining the continuity of such activities.

The very choice of approach to process improvement depends on the objective and specific situation. However, whatever approach is used, processes need to be carefully analysed. And here a detailed visualisation of the processes carried out in the organisation can be helpful. It is important to distinguish the description of processes on the basis of their actual performance from the process diagram described in the procedure, which shows how managers imagine their performance, not how it is introduced in reality. Comparison of the performance of the process actually implemented with its description included in the procedure allows to understand the functioning of the organisation, to identify weak and strong links in the process, including idle actions. Specialised software can be used to visualise organisational processes.

Below, as an example, the process of developing internal regulations which is a sub-process of the identification and introduction of OSH legislation with use of ARIS Toolset software is presented (Figure 2.4) (ARIS Community). ARIS Toolset uses three basic logical operators: "and" (meaning that two or more actions are taken in parallel), "or" (meaning that at least one of given actions has to be taken in parallel) and "xor" (meaning that only one of the two or more actions may be taken in parallel) (Scheer 1999). In the example presented the logical operator "xor" used means that either the draft internal OSH regulations are approved and the process is finished, either they are not approved and it is necessary to repeat some activities. The direction of the process or the information flow is indicated by arrows.

When selecting processes for improvement, apart from the analysis of the process as such, the process is evaluated in terms of:

- effectiveness, i.e. the ability to achieve a goal – effectiveness is usually measured using leading indicators;
- efficiency, which is to be measured by the cost and time of the process; and
- flexibility, i.e. its ability to change quickly.

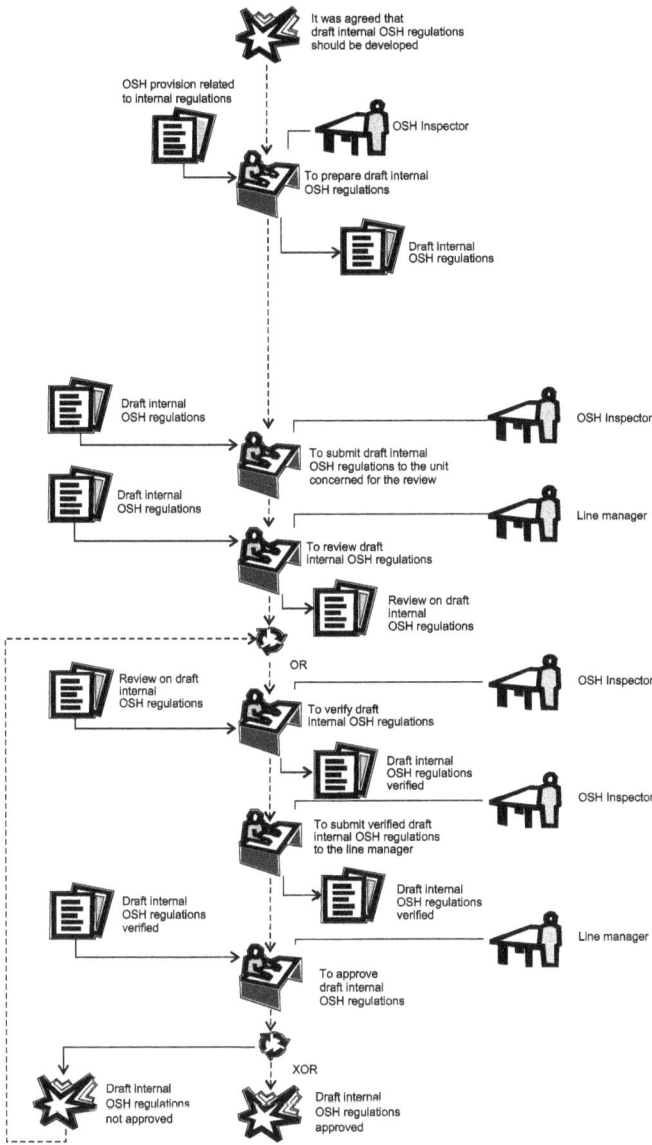

FIGURE 2.4 Example of the process of developing internal regulations (reprinted, with permission, from: Pęciłło 2003, Identyfikacja i modelowanie procesów zarządzania bezpieczeństwem i higieną pracy w przedsiębiorstwie. [Identification and modelling of safety and health management processes in the company] *Bezpieczeństwo Pracy. Nauka i Praktyka* 2:9–12)

The need to measure occupational safety and health management processes is indicated by the ISO 45001 standard (ISO 2018), according to which the criteria for introducing occupational safety and health management processes within an occupational safety and health management system should be defined.

2.2.2 MANAGEMENT OF OSH PROCESSES THROUGH THE APPLICATION OF INDICATORS

2.2.2.1 Development of a Set of Indicators

One of the success factors for a process approach is the measurement of individual processes. This measurement helps to determine whether the operation of these processes is going according to plan and leads to the intended outcomes. Therefore, the measurement of processes is crucial for the continuous improvement thereof and all processes should be monitored with a set of appropriate key performance indicators (KPIs). Knowledge of the current values of these indicators and the trends in the change of these values helps managers to make rational decisions about process control, including determining corrective actions if a process goes wrong (Trkman 2010).

During a process a number of actions are taken, carried out in specific organisational units or areas of an organisation, by persons with appropriate qualifications and authorities. These actions lead to the achieving of specific states and the generation of specific documents, within specific deadlines. States and documents are often subject to approval by line managers.*

All data characterising a process, such as events, states, authorities of persons involved in the process, timeliness of individual actions, correctness of created documents and assessment of their compliance with overriding documents, etc., should be monitored by process owner so that the owner can react on an ongoing basis to possible irregularities in the performance of the process and correct them in order to obtain the intended result (process output). Separate analysis of each of these pieces of data would be time-consuming and give little idea of the process as a whole. This data can be organised and processed collectively, and then made available to the process owner in the form of indicators.

There are many sources available in the literature presenting sets of KPIs to be used by organisations in the area of occupational safety and health. Some of these sets were developed for use in specific industries, such as construction (Poh et al. 2018; Lingard et al. 2017; Dingsdag et al. 2008; Hinze et al. 2013) and mining (Stacey 2012; The Chamber of Minerals & Energy Western Australia 2004; Vinem 2010). Others relate to the measurement of the effectiveness of OSH management systems (Basso et al. 2004; Podgórski 2015; Mohammadfam et al. 2017; Carson and

---

* Explaining this on the example of the audit process: one of the actions carried out within this process is the planning of the audit, for which, according to the audit program, the designated auditor is responsible. The plan is area-specific (e.g. covers operational unit). The taking of this action results in the creation of a new document, i.e. an audit plan and achieving the "audit has been planned" state. The plan should be drawn up within the time limit set out in the audit program and – depending on the organisation's internal regulations – be approved by the management (e.g. an OSH management system representative, process owner, etc.).

Snowden 2010). What the authors unanimously emphasise is the classification of indicators into outcome (lagging, trailing, negative) and leading (positive) (Reiman and Pietikäinen 2012; Lingard et al. 2017, Sinelnikov et al. 2015; Swuste et al. 2016, Amir-Heidari et al. 2017). While there is a widespread perception of the need to measure OSH performance using both groups of indicators, leading indicators are seen as more useful in supporting decision making. It is believed that, inter alia, they make possible proactive monitoring of processes, as they indicate the possibility of non-conformities (Sinelnikov et al. 2015; Leveson 2015) and provide information essential to define effective preventive and corrective actions (Toellner 2001). Therefore, they are sometimes referred to as early warning indicators (Oien et al. 2011). The advantages of leading indicators are related to the fact that they are calculated on the basis of current data characterising the performance of processes. Output indicators, on the other hand, provide information primarily about negative events that have occurred in the past, i.e. what can no longer be changed. They thus only serve as a basis for evaluating past actions and, possibly, drawing conclusions on the effectiveness thereof for the future.

The basic problem related to supporting process management with the use of KPIs is their proper selection. Unfortunately, many organisations make wrong decisions already during this phase of the implementation of indicators: the most important criterion for the selection of indicators becomes the ease of their measurement. As a result, instead of indicators that provide relevant information about the processes, those that can be measured with the least possible effort are chosen. This approach is perfectly summarised in a book by Kaplan and Norton (1996), who write, *"Companies should not fall into the trap of 'if you can't measure what you want, want what you can measure'".*

However, in organisations where this type of distortion of the indicator-based approach occurs, measuring and reporting the value of indicators, instead of supporting process implementation, only disturbs it. Calculating the value of such indicators decreases the effectiveness of work, and, in extreme cases, negatively affects the performance of processes and their outcomes, because instead of focusing on the achievement of process objectives, workers focus on what is being measured (e.g. on timely performing at all costs a given action within the process). The management, who are often accountable for the indicators reaching the expected values, require workers to perform their tasks in a way that ensures this, and having received worthless information about meeting these requirements, they can expect a positive assessment from their superiors This attitude of the management is being consolidated if a system of rewards and penalties for achieving the required values of indicators has been introduced in the enterprise: managers then "learn" to manipulate these values, thus creating a false image of safety in the enterprise (Hale 2009). On the other hand, the need to process easily measurable data that is not linked in any way to the real objectives of the processes causes dissatisfaction and frustration among the workers obliged to measure and calculate the values of the indicators, as they are aware of the unproductivity of their work and the uselessness of its results. This reduces the workers' morale, limits their ability to develop and self-fulfil through work, and takes time which they could use in better ways, e.g. to improve processes.

The organisation's safety and health objectives must be taken into account in the first instance when deciding on the introduction of indicators and when selecting the indicators to be monitored. The indicators should provide information on whether the organisation's processes support the achieving of these objectives and to what extent these objectives have been achieved. Furthermore, the value of the indicator should provide the information needed for decision making. If it is known what kinds of problems and questions recur periodically when supervising the area of occupational safety and health, it is worth ensuring a regular flow of information to answer such questions (Marr 2012).

When developing a set of indicators, one should ensure their relevance to the needs of the organisation. However, in some areas it is also reasonable to use indicators that are commonly used in other organisations, in order to be able to compare results with those of competitors (Marr 2012).

The set of indicators used in the area of occupational safety and health should be updated and continuously harmonised with the changing objectives of individual processes. In addition to the revision of objectives, another stimulus for the introduction of new indicators should be a situation in which the values achieved in the long term by those used so far remain consistently at a high satisfactory level and no longer constitute a basis for process improvement, but only prove that actions taken in the distant past have brought the expected and lasting effect.

The basis for selecting indicators for monitoring in an organisation can be a set of criteria developed by Hale. He argues that a good set of indicators should:

- be valid – actually measure what was to be measured;
- be reliable – individual indicators give the same values in given situations, even if the measurements are made by different people;
- be sensitive – changes in the value of indicators proportionally reflect changes in the measured phenomena;
- be representative – measure all relevant aspects;
- prevent a biased assessment; and
- be cost-effective – the information obtained from it is worth more than the cost of obtaining it (Hale 2009).

Table 2.3 presents examples of leading KPIs described in the literature, which can be used to measure selected processes in OSH management systems.

2.2.2.2 Implementation and Application of KPIs to Monitor OSH Processes

In order to be able to obtain information from the indicator values, the data processing rules necessary to perform the relevant calculations must be developed and implemented. The degree of complexity of these rules and time required to calculate the indicator values depends, inter alia, on the size of the organisation (the more workers and/or sites where particular parts of the enterprise are located, the more difficult it is to obtain and aggregate the necessary data) and on the extent to which IT tools available in the organisation can be used for this purpose.

The first step to calculate the values of selected indicators is to determine mathematical formulae to be used for this purpose. It may seem that most of these formulae

TABLE 2.3
KPIs for measuring operational performance of OSH management system (assigned to ISO 45001 processes)

No.	Process	KPI proposed for a given OSH management system process
1.	Leadership	Worker perception of management leadership and commitment to OSH (rating resulting from the survey)
2.		% of safety-focused actions with demonstrated leadership carried out by managers (against target)
3.		% managers attended OSH leadership training (against total number of managers)
4.		% of audit results reviewed by senior managers
5.		% of incident investigations with managers' involvement (against the total number of incident investigations)
6.	Organisational roles, responsibilities and authorities	% of job descriptions reviewed/assessed and updated for conformity with OSH requirements
7.	Hazard identification	% of hazards with control measures applied (against the total number of identified hazards)
8.		Number of hazards related to particular factors (a type of hazards to be defined) per unit/workgroup
9.		Number of hazard investigations conducted for non-routine operations by means of Job Safety Analysis (JSA) per organisational unit
10.		% of JSA-based hazard investigations reviewed (against the total number of JSAs)
11.	Assessment of OSH risks	Number or % of risk assessments completed or reviewed (against planned)
12.		% of risk assessments conducted indicating high level risks (against the total number of risk assessments)
13.		% or number of risk assessments performed on new plants/processes and prioritised with control measures
14.		% of incident investigations which caused changes of risk assessment results (against the total number of incident investigations)
15.	Planning to take action	% of risk control measures or risk assessment-based corrective actions successfully completed (against planned)
16.		Average time required for getting particular hazards and risks under control (reducing to an acceptable level) once they have been identified and assessed
17.		% of reduction in exposure for hazardous activities (in hours of exposure or number of workers exposed)
18.	Planning to achieve OSH objectives	% of sites/departments that have determined a plan of OSH-related activities
19.		% of OSH objectives achieved according to the plan

(Continued)

TABLE 2.3 (CONTINUED)

KPIs for measuring operational performance of OSH management system (assigned to ISO 45001 processes)

No.	Process	KPI proposed for a given OSH management system process
20.	Competence	% of permits to work reviewed and positively assessed (against the total number of permits required for specific jobs subject to periodic reviews)
21.		% of OSH training courses completed as per plan
22.		Average number of hours for OSH training per person
23.		% of workers trained in accordance with the planned OSH training programme
24.		OSH training effectiveness and appropriateness (based on findings from survey of trainees)
25.		Number or % of workers that have completed required first aid training
26.		% of incidents in which training was identified as a major contributor
27.	Awareness	Number of failures and cases of breaking safety rules reported by workers, per unit
28.		The level of workers' involvement in OSH improvement activities (safety culture level)
29.		Number of near-misses reported by workers (e.g. per 10 workers)
30.	Information	Number of OSH management communications (announcements) distributed among workers via the intranet
31.		Number of OSH-related announcements published in the enterprise's newspapers
32.	Communication	Number of meetings organised between senior managers and workers at which OSH management issues were discussed
33.		Number of safety meetings held before commencing particularly dangerous jobs, against number of new cases when such jobs are being taken up
34.		% of OSH committee meetings held as per plan
35.	Participation and	% of workers participating in competitions on OSH knowledge
36.	consultation	% of workers involved in making suggestions for OSH improvements (each worker counted once irrespective of a number of suggestions made)
37.		Number of OSH inspections carried by representatives of trade unions
38.		% of OSH improvement proposals accepted by managers for introduction

(Continued)

TABLE 2.3 (CONTINUED)

KPIs for measuring operational performance of OSH management system (assigned to ISO 45001 processes)

No.	Process	KPI proposed for a given OSH management system process
39.	Operational planning and control – General	Number or % of preventive maintenance activities performed according to schedule
40.		Number of OSH inspections
41.		Number of non-conformities detected and reported during OSH inspections
42.		Number of management safety visits compared with the number planned
43.		Cost of equipment failures
44.	Procurement	% of service and purchase contracts with OSH clauses
45.		% of specifications with reference to OSH requirements when purchasing equipment or contracting service
46.		Number of risk assessments carried out prior to the introduction of new materials and equipment into the workplace
47.	Contractors	Number of OSH management audits carried out in contractor's plants
48.		Number of joint meetings with contractors on OSH issues
49.	Emergency preparedness and response	% of workers trained in emergency response and co-ordination per unit/plant/shift
50.		% of emergency systems and first aid equipment inspected according to schedule
51.		% of action items closed out within given timeframe following emergency response debrief
52.	Monitoring, measurement, analysis and evaluation –	Number of completed vs. planned monitoring activities (e.g. safety behaviour observations, measurements of working environment parameters, etc.)
53.	General	Number of managers that have reviewed their departments' OSH performance
54.	Internal audit process	% of internal audits conducted to schedule
55.		Number or % of workers that have received required training for OSH management system auditing
56.		Number of sub-standard conditions identified as a result of OSH management system internal audits
57.		% of workers taking part in internal OSH management system audits
58.	Management review	% of OSH management system components and activities reviewed according to schedule
59.		Number of management meetings dedicated to OSH management system reviews

(Continued)

TABLE 2.3 (CONTINUED)
KPIs for measuring operational performance of OSH management system (assigned to ISO 45001 processes)

No.	Process	KPI proposed for a given OSH management system process
60.	Incident, nonconformity and corrective action	% of total number of corrective/preventive actions completed on time according to the action programme
61.		% of corrective and preventive actions from incident investigations completed on time according to the action programme
62.		% of corrective and preventive actions completed from safety climate/cultural survey
63.		Number of workers participating in investigation teams
64.	Continual improvement	% of change in the overall rating of OSH performance over a specified timeframe
65.		% of monitoring results (e.g. safety behaviour observations, measurements of working environment parameters, etc.) that are above or below the target values against the total no. of monitoring results

Adapted from working materials developed within the project titled *Development and Validation of a KPI-Based Method and a User-Friendly Software Tool for Resilience-Focused Measurement of OSH Management System Performance*, which was carried out by CIOP-PIB (Central Institute for Labour Protection – National Research Institute, Poland), FIOH (the Finnish Institute for Occupational Health) and Tecnalia Research & Innovation, Spain.
The project was funded within the SAF∈RA Network (www.safera.eu) by CIOP-PIB, Instituto Vasco de Seguridad y Salud Laborales (OSALAN), Basque Country, Spain, FIOH, and the Finnish Ministry of Social and Health Affairs.

are simple quotients or quotients of the sums, but it is important to carefully define the numerators and denominators of these quotients and, if necessary, the form of expression of the results (e.g. in percentages, per 10 workers, etc.). The numerator and denominator may be values that describe the number of events and actions that were planned or performed, as well as people who performed or should have performed those actions. Clarification is needed as to what type of event or activity meets the criteria to be included in the numerator/denominator value (e.g. in the case of indicators relating to safety and health training, indicate whether it is initial, periodic or other training). If the value of a numerator or denominator means the number of persons, it should be specified which persons are concerned (management, workers exposed to certain factors, workers of the selected organisational units, all workers, all workers plus other persons present on the organisation premises, etc.).

The data needed to calculate the value of the indicators and the place(s) of their generation are then identified. As individual data records can be derived from individual persons (e.g. the auditor who developed the audit plan), every effort should be made to ensure comprehensive aggregation of this data and to provide workers responsible for providing the data with appropriate tools to support this process (e.g. to provide space in the enterprise intranet for reporting the elaboration of the audit plan). The source of data used to calculate the indicators may also include reports made by various people inside and outside the enterprise and results of measurements (e.g. measurements of concentrations and intensities of harmful, hazardous and noxious factors in the working environment). In this case, it should be determined who will be responsible for providing the relevant data to the person calculating the value of the given indicator, in what form and within what timeframe.

Some of the data necessary for the calculation of indicator values can be generated or saved in the IT systems and databases used in an enterprise, which greatly facilitates the sharing and further processing thereof to calculate these values. It is also worthwhile to analyse the possibility of an expansion of IT infrastructure that will facilitate the collection and processing of necessary data.

One of the important elements of the introduction of the indicator approach in process management is to decide to whom – in addition to the process owner – to provide information about the calculated values. The group or groups of these persons (users), e.g. management, workers, etc., should be specified. The communication channels through which the indicator values will be presented to the individual groups should also be clarified. Management may receive them in the form of a report, e.g. as input data for the management review, workers of the enterprise may learn about these values from the enterprise newsletter. If possible, it is advisable providing the process owner with access to the value of the indicator updated on an ongoing basis, by processing the relevant data in the IT system.

In order to ensure that trends in the evolution of indicator values are detected, indicators should be calculated at fixed intervals. The frequency of these calculations is determined by many factors, such as the variability of the indicator value over time or the importance of this value for the decisions made in the enterprise on a daily basis. For some indicators, e.g. those calculated on the basis of surveys, a lower measurement frequency is also determined by its high cost.

Process management can be supported by setting reference values for individual indicators, i.e. by indicating which ranges of values confirm the correct operation of a given process and which indicate that the process is not going well. Relating the current value to a reference value makes possible the immediate identification of processes the functioning of which needs to be improved and can provide a basis for prioritising corrective actions in the OSH management system. This involves deciding how to present the indicators, e.g. numerical, graphical or tabular. Colour policy can also be used for this purpose, i.e. the value of the indicator can be presented with the use of such a colour, which – according to the rules adopted in the given organisation – will indicate to what extent this value is in line with expectations. The unification of colour policy for all indicators used on an organisation scale may be helpful primarily for the top management, which, thanks to such an approach, will be able to identify, "at a glance", from among the numerous processes carried out in the enterprise those the operation of which is proceeding with disruption.

Figure 2.5 is a diagram of the introduction of indicators to support the management of OSH processes by steps. The names of those steps that are not necessary but can improve the monitoring of indicator values and their use for decision making are in italics.

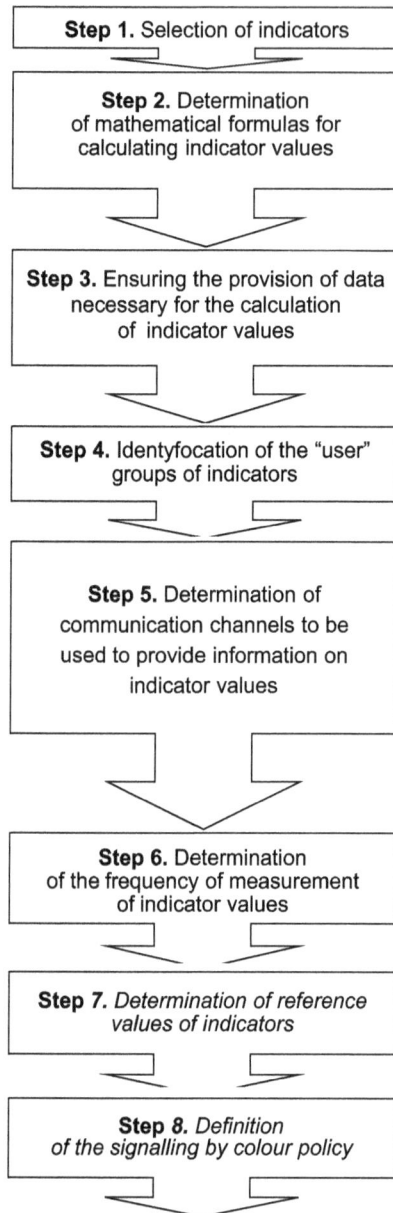

Step 1. Selection of indicators

Step 2. Determination
of mathematical formulas for
calculating indicator values

Step 3. Ensuring the provision of data
necessary for the calculation
of indicator values

Step 4. Identyfocation of the "user"
groups of indicators

Step 5. Determination of
communication channels to be
used to provide information on
indicator values

Step 6. Determination
of the frequency of measurement
of indicator values

Step 7. *Determination of reference
values of indicators*

Step 8. *Definition
of the signalling by colour policy*

FIGURE 2.5   Diagram of introduction of indicators to support the process approach

The description of the methodology for the selection and introduction of indicators to support the management of OSH processes proves that the use of indicators is a time-consuming task and requires systematic work. However, the indicator
approach is not only about determining measurable results and regularly calculating
the value of indicators. Its use is only justified if the information thus obtained can
be applied to improve the safety of workers.

First of all, it is worth remembering that the values of indicators illustrate the state of processes "in a nutshell", offer their quick diagnosis and are a source of highly aggregated information. They are ideal for use during meetings with top management to briefly discuss current problems or during meetings with workers – to boast about achievements, inform people about challenges, and draw attention to difficulties in OSH management. However, for the owner of an occupational safety and health process, the value of an indicator should only be a prelude to a more thorough analysis of the process at regular intervals.

The first benchmark for an assessment of the calculated value of an indicator is, of course, its reference value determined during the implementation phase. If the measured value is negative (i.e. falls within a range which means that the process is not working properly), the circumstances of the situation must first be explained and the causes identified. As indicated previously, the indicator captures the picture of the process in a certain point of its performance and it is necessary to examine the entire process in order to explain the value of the indicator which is not in line with expectations. Such a value should be compiled with the values recorded in previous measurements – the trend in value changes should be determined and any events and circumstances that may have influenced its shape should be linked to this trend. At the same time, it is worth analysing the then current situation: first, in terms of its possible negative consequences; and, second, in terms of the actions that need to be taken in order for the process to operate smoothly again.

If the measured value is positive (i.e. falls within a range which means that the process is working properly), it is worth considering what success factors helped to achieve this value. The identification of such factors is particularly important if a negative value was recorded during a previous measurement of an indicator. In such a case, the actions taken between measurements should be carefully analysed in order to identify any actions that support the process. Just as recording a negative value of an indicator, also recording a positive value should be an incentive to analyse the trend of changes in these values. It may turn out that, although positive, the indicator value is approaching the threshold of the reference value and may soon exceed it. In addition, as previously mentioned, long-term recording of positive values of indicators may mean that the benchmark has been set too low and it is high time to verify the reference value or even to consider changing the indicator.

What is known as the DIKW pyramid (data – information – knowledge – wisdom) can help to clarify the role of indicators in management (Ackoff 1989; Rowley 2007). This concept assumes that wisdom (located at the top of the pyramid) is the ability to improve effectiveness, in turn knowledge makes it possible to transform the information one possesses into instructions, while the latter are created on the basis of available data (constituting the lowest level of the pyramid). The concept states that data, information and knowledge, although necessary to acquire the wisdom that enables effective management, do not guarantee it. Moving to a higher level of the pyramid requires adequate processing of resources at lower levels.

The indicators, being sources of information about processes, are at the second level from the bottom of the pyramid (Figure 2.6). When interpreting the concept of Ackoff (1989) in the context of supporting OSH process management with the use of indicators, it can be concluded that knowing the value of an indicator is only useful information for the process owner if they also know where this value comes from,

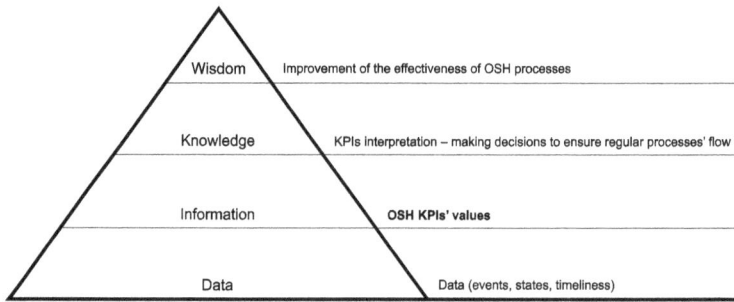

FIGURE 2.6 The use of OSH KPIs in decision making (adapted from Rowley [2007])

which in turn enables them to define actions to ensure that the process runs properly in the future. A wise process owner, using the values of the indicators, will be able, based on their knowledge, to improve the effectiveness of the process. In other words, knowing the value of indicators is not useful for supporting OSH management if the manager does not understand how the process performance has led to the fact that one value of indicator has been recorded rather than another.

OSH process management can be successfully supported by using appropriate indicators to measure it and make decisions based on the results obtained. However, the application of this approach must not come down to ensuring that the indicators achieve their intended values and end when the objective has been achieved. The use of indicators can be a successful idea if they are properly selected and introduced, calculated at regular intervals and communicated to the relevant persons who, using the knowledge and wisdom gained, manage the further performance of the measured processes. Therefore, it is worthwhile to respect the following rules when using indicators:

- keep in mind that your overarching goal is to prevent accidents and occupational diseases;
- read the information behind KPI values instead of putting your faith in numbers; and
- interpret KPI values to make decisions rather than simply order people to achieve KPI-based goals.

2.2.3 ACTIVITY-BASED COSTING METHOD AS A TOOL FOR OSH-RELATED PROCESS OPTIMISATION

Analysis of the costs of processes carried out in the field of an enterprise's occupational safety and health makes monitoring and evaluation of actions improving the performance of these processes possible. The method that enables such an economic analysis is the Activity-Based Costing (ABC), which, unlike traditional methods of cost calculation in organisations, making it possible to determine costs by departments and by type, focuses on activities as causes of costs. First of all, the analysed process has to be described in detail at the level of activities implemented in this process (see Figure 2.4) and for these activities individual costs have to be determined.

Then the costs for the entire process implemented on a one-off basis have to be calculated. Finally, the number of times the process is carried out within a specific time period (e.g. year) has to be assessed and the total costs of the process calculated (Miller et al. 2000). Activity-Based Costing makes it possible to precisely determine the costs of individual activities and entire economic processes in the organisation, and thus enables to determine which activities generate the highest costs and who is responsible for incurring these costs.

In order to carry out an economic analysis of the organisational process, it is therefore necessary to:

- identify activities within the organisation and model the process;
- identify cost drivers, i.e. factors causing a change in the costs of the activities, e.g. in a risk assessment process, it may be the number of jobs covered by the activities involving monitored. One activity may involves several cost drivers;
- identify the main cost-generating units and assign them to individual activities. It should be remembered that these units do not always overlap with organisational units operating in a formal organisational structure. For example, a risk assessment team is not provided for in an organisational structure (it may have a permanent composition or be appointed on an ad hoc basis each time there is a need to carry out a risk assessment), but it is the unit responsible for generating costs of activities in the risk assessment process. It should be noted here that more than one unit may participate in the costs of individual activities;
- identify and prioritise costs in the organisation and then assign them to individual activities. In organisational processes, these are primarily personnel costs. This means that for each cost-generating unit in each activity, the time of its work should be determined and the cost of the activity should be calculated based on an hourly rate. Likewise, the costs of the materials used are dealt with. Fixed costs for the economic analysis of OSH management processes should preferably be omitted.

It is important to remember that monitoring of costs of processes with the activity-based costing method requires that each time entered data is real and not imagined by or desired from the point of view of the organisation's management. Therefore the cost analysis requires ongoing consultation with process leaders.

Thus, the economic analysis of organisational processes using the ABC method is based on precise data, so it is time-consuming. Therefore, before carrying out such analyses, it is necessary to select processes for which cost calculations will be carried out. These should first be processes to be improved (or being subject of other planned changes). The results of the cost analysis provide information on the costs of the processes before and after changes. An example of using the ABC method for estimating savings resulting from process improvement in various organisations is presented in Table 2.4. Process improvements were rather of incremental nature, based on the concept of continuous improvement in the spirit of the TQM concept. Regardless of the organisation and type of process, the improvement of processes

TABLE 2.4

Costs of process performance before and after optimisation (improvement) in the enterprises studied (Adapted from Pęciłło 2005a)

Enterprise	Costs before optimisation in EUR	Costs after optimisation in EUR	Savings in %
	Trainings in OSH		
chemical	1,307	no optimisation	n/a
pharmaceutical	1,781	1,482	17%
furniture	680	665	2%
construction	582	no optimisation	n/a
metallurgical	499	no optimisation	n/a
	Hazard identification and occupational risk assessment		
chemical	4,960	4,473	10%
pharmaceutical	7,036	6,038	14%
furniture	3,505	3,274	7%
construction	3,116	2,379	24%
metallurgical	2,522	2,315	8%
	Monitoring working conditions		
chemical	5,368	4,507	16%
pharmaceutical	5,586	5,544	1%
furniture	2,750	no optimisation	n/a
construction	2,255	1,946	14%
metallurgical	1,571	1,504	4%

resulted mainly in "giving more authorisation for the safety officer and line managers, simplifying process performance through deleting repeated and unnecessary activities (activities that do not influence the level of meeting process objectives) and assigning clearly activities performed within the processes studied to one organisational unit or one post" (Pęciłło 2005a).

The second application of the ABC method is the ability to estimate savings as a result of the integration of OSH management processes with other processes. An example of such analyses for selected training, internal communication, internal audits and identification of legal and other requirements and the introduction thereof into the practice of an enterprise is presented in Table 2.5.

Thanks to the integration of management systems an enterprise can gain apart from economic benefits also organisational ones, for example eliminating a number of unnecessary activities conducted in the organisation, reducing a number of documents developed in the organisation, improving a flow of information and a flow of documents and improving communication between organisational units (Pęciłło 2005b).

Economic analysis of the costs of OSH management processes enables one to estimate the economic benefits resulting from the improvement of these processes or

TABLE 2.5

Cost of processes in non-integrated management systems and savings from integration (in euro)

Process	Cost of processes in two non-integrated management systems (per year)	Savings (per year)		Cost of processes in three non-integrated management systems (per year)	Savings (per year)	
Identification of legal requirements	1,353	639	47 %	2,047	1,225	60 %
Training	7,352	1,424	19 %	10,674	3,499	33 %
Auditing	3,696	1,338	36 %	5,553	2,725	49 %
Internal communication	2,336	932	40 %	3,504	1,864	53 %
Total	14,737	4,333	29 %	21,778	9,313	43 %

from their integration with other processes carried out in the enterprise, for example with those related to quality assurance or environmental protection. Therefore it enables one to assess the advisability of improvement measures. However, it must be remembered that economic benefits are only one type of benefit. The improvement of the process of document circulation and communication, limitation of the number of documents and activities performed, and unambiguous assignment of tasks and responsibilities to organisational units, which in effect leads to improved communication between them are of the same importance.

2.3 SUMMARY

To increase the effectiveness of management systems, organisations primarily focused on strategic processes that directly created added value. Less attention was usually paid to auxiliary processes including OSH-related ones. The situation was changed recently as the new standard ISO 45001:2018 "Occupational Safety and Health Management Systems – Requirements with Guidance for Use" has been adopted and published, in which process-oriented approach is the basic prerequisite for the OSH management system compliance with this standard.

The process perspective enables managing of OSH-related processes more effectively. The starting point for the introduction of the process approach is identification and modelling, i.e. graphical representation of organisational processes. The level of detail of such processes depends on the specificity of each organisation, its organisational structure, management style, and the purpose of introducing process management principles. It should be remembered that processes cannot be equated with procedures. The former reflect reality, the latter are the ideas of the work being done, often in the most efficient way from the point of view of the creators of procedures. Thus the very modelling of processes enables one to verify the effectiveness of management by contrasting an organisation's documentation with reality.

Process management includes process improvement by identifying weaknesses in a process, e.g. idle or redundant activities from the point of view of achieving the assumed process objectives and then changing the performance of processes. In addition, process management requires monitoring of their effectiveness and efficiency.

The effectiveness of processes, understood as the ability to achieve specific objectives, is measured by a set of indicators. It is advisable that the indicators not only measure the achievement of objectives but also identify process dysfunctions. Therefore it is necessary to establish not only output indicators, but above all leading indicators to diagnose the problem before a work accident or other adverse event occurs. It should also be remembered that the indicators themselves are insufficient to assess the functioning of organisational processes, as in many cases they are based on non-measurable data and do not really provide objective data. Furthermore, in many cases it is difficult to set objective criteria for their assessment, especially in the rapidly changing internal and external environment of the organisation. Therefore, the indicative assessment requires expert support in the final phase.

The measuring of the effectiveness of the processes, especially the processes of OSH management, understood as the ratio of the effects to the expenditures incurred, can be approached in a simplified way and reduced to monitoring the costs or time of implementation of these processes. Calculation of process costs using the activity-based costing method allows precise determination of process costs, and thus proves to be useful when it is necessary to estimate savings resulting from process improvement or their integration with other processes.

The introduction of the process approach to OSH management is also the starting point for strategic management in this area using the Balanced Scorecard concept.

REFERENCES

Ackoff, R. L. 1989. From data to wisdom. *Journal of Applied Systems Analysis* 15:3–9.
Amir-Heidari, P., R. Maknoon, B. Taheri, and M. Bazyari. 2017. A new framework for HSE performance measurement and monitoring. *Safety Science* 100:157–167. https://reader.elsevier.com/reader/sd/pii/S0925753516304350?token=D1532929E0B5AB6B8B1 AE3E59AE3649F5404C7DDAB1155C0206CBDD4415C2ABA85D6DDEA403A36C A8FD6176F3E851CF3 (accessed September 19, 2019).
ARIS Community. https://www.ariscommunity.com/ (accessed January 17, 2020).
Basso, B., C. Carpegna, C. Dibitonto, G. Gaido, A. Robotto, and C. Zonato. 2004. Reviewing the safety management system by incident investigation and performance indicators. *Journal of Loss Prevention in the Process Industries* 17(3):225–231. https://reader.elsevier.com/reader/sd/pii/S0950423004000051?token=C2834DC80247E01DD304D52A 7B91440CE2D23AF2867F2CE4CA81FFF2CF96301F94387FC2F0E0FF96CB58 6699F842F507 (accessed September 19, 2019).
vom Brocke, J., and M. Rosemann (edit). 2015. *Handbook on Business Process Management 1 Introduction, Methods, and Information Systems.* 2nd edition. Springer-Verlag: Berlin Heidelberg.
Carson, P. A., and D. Snowden. 2010. Health, safety and environment metrics in loss prevention – part 1. *Loss Prevention Bulletin* 212(April):11–15. http://web.b.ebscohost.com/ehost/pdfviewer/pdfviewer?vid=1&sid=a043a845-5e51-49f8-84b0-a3b6120eab29% 40pdc-v-sessmgr03 (accessed September 19, 2019).
Davenport, T. H., and J. E. Short. 1990. The new industrial engineering: Information Technology and business process redesign. *Sloan Management Review* 4:11–27.

Davenport, T. H. 1993a. Need radical innovation and continuous improvement? Integrate process reengineering and TQM. *Planning Review* 3(3):6–12.

Davenport, T. H. 1993b. *Process Innovation: Reengineering Work through Information Technology.* Harvard Business School Press: Boston, MA.

Deming, W. E. 1986. *Out of Crises: Quality, Productivity and Competitive Position.* Cambridge University Press: Cambridge, MA.

Dingsdag, D. P., H. C. Biggs, and D. Cipolla. 2008. Safety Effectiveness Indicators (SEI's): Measuring construction industry safety performance. In: *Proceedings of the Third International Conference of the Cooperative Research Centre (CRC) for Construction Innovation – Clients Driving Innovation: Benefiting From Innovation*, ed. J. Thomas, and A. Piekkala-Fletcher. Gold Coast, Australia. https://eprints.qut.edu.au/15438/1/15438.pdf (accessed September 19, 2019).

Gage, S. 1993. The management of key enterprise processes. *Centre Quality of Management Journal* 2:13–18.

Hale, A. 2009. Why safety performance indicators? *Safety Science* 47(4):479–480. https://reader.elsevier.com/reader/sd/pii/S0925753508001227?token=308EDE1CCE0F661F2474C1511F6EEE50067AFB2DE7BE7EA906CAB2B26AA9637D24D9097B4DBEF9C847D1CD9F9D8EE919. (accessed September 19, 2019).

Hammer, M. 1990. Reengineering work: Don't automate obliterate. *Harvard Business Review* 4:104–112.

Harry, M., and R. Schroeder. 2006. *Six Sigma: The Breakthrough Management Strategy. Revolutionizing the World's Top Corporations.* Crown Business: New York, Currency.

Hinze, J., S. Thurman, and A. Wehle. 2013. Leading indicators of construction safety performance. *Safety Science* 51(1):23–28. https://reader.elsevier.com/reader/sd/pii/S0925753512001361?token=8BB14969A4EC34FC4AB31425DAC972887F63B464238BC834CFF93C6372DA72BDE5BBAB13B23670B5D4FCB2A643E94351. (accessed September 19, 2019).

ISO (International Organization for Standardization). 2018. *ISO 45001:2018 Occupational Health and Safety Management Systems: Requirements with Guidance for Use.* International Organization for Standarization. Geneva, Switzerland.

Ishikava, K. 1985. *What Is Total Quality Control? The Japanese Way.* Prentice-Hall: Englewood Clifs, NJ.

Jeston, J., and J. Nelis. 2014. *Business Process Management Practical guidelines to successful implementations.* 3rd edition. Routledge: New York.

Juran, J. M. 1989. *Juran on Leadership for Quality.* Free Press: New York.

Kaplan, R. B., and L. Murdock. 1991. Core process redesign. *The McKinsey Quarterly* 2:27–43.

Kaplan, R. S., and D. P. Norton. 1996. *Translating Strategy into Action.* The Balanced Scorecard, Harvard Business School Press: Boston, MA.

Keen, P. G. W. 1997. *The Process Edge.* Harvard Business School Press: Boston, MA.

Kueng, P. 2000. Process performance measurement system: A tool to support process-based organizations. *Total Quality Management* 11(1):67–85.

Leveson, N. 2015. A systems approach to risk management through leading safety indicators. *Reliability Engineering and System Safety* 136:17–34. https://reader.elsevier.com/reader/sd/pii/S0951832014002488?token=F7BAD7D792BF5C8022B5C6C4CDE628346B44E17EA79C40C056F1C2E26D569C614CD7EF13427DD34769C18E6492F2EAE3 (accessed September 19, 2019).

Lingard, H., M. Hallowell, R. Salas, and P. Pirzadeh. 2017. Leading or lagging? Temporal analysis of safety indicators on a large infrastructure construction project. *Safety Science* 91:206–220. https://reader.elsevier.com/reader/sd/pii/S0925753516301837?token=CD678691A01847EA2FC7ADE5FE21BE3324E0CCAC5A5682429BDCB2DBF5B0AB041C635A9E6C007E89B231BC2145164C3F (accessed September 19, 2019).

Marr, B. 2012. *Key Performance Indicators (KPI): The 75 Measures Every Manager Needs to Know.* FT Press: Harlow, UK.

Melin, U. 1999. The process perspective in Total Quality Management and business process reengineering – A critical review. In: *Proceedings of the International Conference on TQM and Human Factors – Towards of Successful Integration*, ed. J. Axelsson, B. Bergman, and J. Eklund. Eds. Centre for Studies of Humans, Technology and Organisation. Linkoping, Sweden. 253-260.

Miller, J. A., K. Pniewski, and M. Polakowski. 2000. *Zarządzanie kosztami działań.* (Activity-Based Management). WIG-Press: Warszawa.

Mohammadfam, I., M. Kamalinia, M. Momeni, R. Golmohammadi, Y. Hamidi, and A. Soltanian. 2017. Evaluation of the quality of Occupational Health and safety management systems based on key performance indicators in certified organizations. *Safety and Health at Work* 8(2):156–161. https://reader.elsevier.com/reader/sd/pii/S2093791116300634?token=5047C22E83993D21AD296B4FA7A0C7AF837DBB7E3865057045B7E6703B1CD5C1BA513F1611B6C7B5FBB9D5A47B9BA7CD (accessed September 19, 2019).

Oien, K., I. B.Utne, and I. A. Herrera. 2011. Building Safety indicators: Part 1 – Theoretical foundation. *Safety Science* 49(2):148–161. https://reader.elsevier.com/reader/sd/pii/S0925753510001335?token=86BCE7944FBAAAFF047E9BA95B47EACF2C970FD97AAD8A880786A7EF598DE8A44266C20B981D8D842683F69A98EE10EF (accessed September 19, 2019).

Pęciłło, M. 2003. Identyfikacja i modelowanie procesów zarządzania bezpieczeństwem i higieną pracy w przedsiębiorstwie. (Identification and modelling of safety and health management processes in the company). *Bezpieczeństwo Pracy – Nauka i Praktyka* 2(379):9–12. http://archiwum.ciop.pl/5958 (accessed September 19, 2019).

Pęciłło, M. 2005a. Optimisation of OSH-related management processes in enterprises. In: *Advances in Safety and Reliability – ESREL 2005, Two Volume Set: Proceedings of the European Safety and Reliability Conference, ESREL 2005, Tri City, Poland, 27–30 June 2005*, eds. K. Kolowrocki. London. A.A. Balkema Publisher. 1541–1546.

Pęciłło, M. 2005b. Usprawnianie procesów zarządzania bezpieczeństwem i higieną pracy w przedsiębiorstwie. (Improvement of safety and health management processes in the company). *Bezpieczeństwo Pracy – Nauka i Praktyka* 1:11–13 http://archiwum.ciop.pl/14100 (accessed September 19, 2019).

Pawłowska, Z., and M. Pęciłło. 2018. *Doskonalenie zarządzania bezpieczeństwem i higieną pracy z uwzględnieniem wymagań i wytycznych normy międzynarodowej ISO 45001.* (Improving occupational safety and health management with regard to the requirements and guidelines of international standard ISO 45001). Warsaw. CIOP-PIB.

Podgórski, D. 2015. Measuring operational performance of OSH management system – A demonstration of AHP-based selection of leading key performance indicators. *Safety Science* 73:146–166. https://reader.elsevier.com/reader/sd/pii/S0925753514003063?token=2C86D2F1C7F174C5C7885724BED794B85B901951E297B36F679B7BFA6C0451F1A975C0C271FDB66F6BA9C8C99CF44471 (accessed September 19, 2019).

Poh, C. Q. X., C. U. Ubeynarayana, and Y. M. Goh. 2018. Safety leading indicators for construction sites: A machine learning approach. *Automation in Construction* 93:375–386.

Reiman, T., and E. Pietikäinen. 2012. Leading indicators of system safety – Monitoring and driving the organizational safety potential. *Safety Science* 50(10):1993–2000. https://reader.elsevier.com/reader/sd/pii/S0925753511001688?token=ABFFCC75F8ED09A750107827C7CD1F787A4E93DC0D0823794A4CED745CCAABEABC03B2DA88A7B817FCD6DB5C9FBBE18D (accessed September 19, 2019).

Rowley, J. 2007. The wisdom hierarchy: Representations of the DIKW hierarchy. *Journal of Information Science* 33(2):163–180. https://pdfs.semanticscholar.org/088d/6a1fa59a8840ab0dff0f2e06d1c1fd7d4012.pdf (accessed September 19, 2019).

Rummler, G. A., and A. P. Brache. 1995. *Improving Performance: How to Manage the White Space on the Organization Chart.* Jossey-Bass: San Francisco, CA.

Scheer, A.-W. 1999. *ARIS – Business Process Modeling.* Springer Verlag: Berlin.

Sinelnikov, S., J. Inouye, and S. Kerper. 2015. Using leading indicators to measure occupational health and safety performance. *Safety Science* 72:240–248. https://reader.elsevier.com/reader/sd/pii/S0925753514002203?token=4DB341AD5FBD3834C79337E27C47272C11D38F90281132270BFFB0C77C507A9B7CF670B9A22D45B88D429BCE238C61FA (accessed September 19, 2019).

Stacey, J. 2012. Overview of leading indicators for occupational health and safety in mining, International Council on Mining and Metals. https://www.icmm.com/website/publications/pdfs/health-and-safety/4800.pdf (accessed September 19, 2019).

Swuste, P., J. Theunissen, P. Schmitz, G. Reniers, and P. Blokland. 2016. Process safety indicators, a review of literature. *Journal of Loss Prevention in the Process Industries* 40:162–173. https://reader.elsevier.com/reader/sd/pii/S095042301530098X?token=7CFD2774ED278D0FC3DF849AB583DC6B547949312311BA13D210C68349C0D147A17A276822A6010CE9056F4A61243227 (accessed September 19, 2019).

The Chamber of Minerals & Energy Western Australia. 2004. *Guide to Positive Performance Measurement in the Western Australian Minerals and Resources Industry.* The Chamber of Minerals & Energy: Western Australia.

Toellner, J. 2001. Improving safety and health performance: Identifying and measuring leading indicators. *Professional Safety* 46(9):42–47. https://cdn2.hubspot.net/hubfs/2176045/Predictivesafety%20December2017/PDF/Improving+Safety.pdf?t=1519044883737 (accessed September 19, 2019).

Trkman, P. 2010. The critical success factors of business process management. *International Journal of Information Management* 30(2):125–134.

Twaróg, J. 2002. Tworzenie Struktury Procesowej. (Establishing a process structure). *Problemy Jakości* 11:13–22.

Vinem, J. E. 2010. Risk indicators for major hazards on offshore installations. *Safety Science* 48(6):770–787. https://reader.elsevier.com/reader/sd/pii/S092575351000055X?token=BAFA22DB64BE55716A1A93043CE67F604426680AF0374AAB6EC7B33950894A9918C95E5A209F97FDB897ABA26BA4500D (accessed September 19, 2019).

3 Application of Fuzzy Cognitive Maps for Modelling OSH Management Systems

Anna Skład

CONTENTS

3.1 FUZZY COGNITIVE MAPS AS A RESEARCH METHOD BASED ON PRACTICAL AND THEORETICAL EXPERT KNOWLEDGE

3.1.1 COMMONLY USED TOOLS SUPPORTING DECISION MAKING IN OSH MANAGEMENT

Fuzzy logic is not logic that is fuzzy, but logic that is used to describe fuzziness. Fuzzy logic is the theory of fuzzy sets, sets that calibrate vagueness (Husain et al. 2017). Fuzzy logic is often associated with sophisticated mathematics that are difficult, if not impossible, to be used by people with general mathematical knowledge. Few people know that the intention of the creators of fuzzy logic was to make it applicable exactly in those fields where numbers and complex mathematical records were unnecessary. The aim of this study is to introduce fuzzy logic and present the possibilities of its application in the enterprise management process, and, more specifically, in the field of occupational safety and health (OSH) management. It is suggested to use for this purpose the fuzzy logic-based method known as fuzzy cognitive maps (FCM).

For effective management, it is necessary to simultaneously analyse the processes executed in the given enterprise as well as external factors and processes that directly and indirectly affect them. Only a comprehensive view of the entire organisation and its context will enable effective planning of activities, including anticipation of problems, identification of opportunities, and reacting in crisis situations. Taking action based on fragmented knowledge about an enterprise may lead to unintentional results. Affecting an individual process without knowing which other factors influence it and what the consequences would be may lead to uncontrollable changes in this process and in the entire management system.

Considering the above, one should ask a question about the possible methods of gathering and processing knowledge which is needed to manage an enterprise. One of them may be the application of key performance indicators (KPIs) which enable the summarising of the most important trends of processes and phenomena occurring in the enterprise and the enterprise context. However, to calculate the values of KPIs it is often necessary to regularly collect and process information from different sources, which is both time-consuming and costly. Furthermore, the value of a KPI needs to be compared over determined periods of time to make it possible to pick out trends. Therefore, having once decided on a certain set of KPIs, an enterprise needs to consequently keep on calculating them as long as it wishes to keep track. Obviously even small change in KPI definition might interfere with conclusions regarding trends and thus make the information obtained less useful. On the other hand, the dynamic environment in which enterprises operate constantly enforces the updating of indicators. Unfortunately, development and introduction of processes to calculate updated measures sometimes takes longer than the maximum available time for making the decision.

Another valuable sources of knowledge about enterprises, on the basis of which management decisions are made, may be conclusions and recommendations from audits, both those carried out by external institutions and those performed internally by enterprises. Usually presented in audit reports, such conclusions and recommendations are subject to analysis, as a result of which actions are taken relating to specific recommendations. Auditors investigate various aspects of an enterprise's operation and present their findings to the managers in charge. This is a stimulus to develop a plan of recovery from non-conformity, failure or crisis situations which prevents or breaks the chain of adverse events. However, a plan of remedial actions prepared under pressure from an auditor, unless thoroughly discussed with all interested parties, although aiming to improve the way the given field works, can in fact influence (not necessarily positively) other fields and change them in an unpredictable and unintended way.

A comprehensive look at an enterprise's operation might be achieved by mapping of business processes using standard notation, usually BPMN. The process map is in this case a diagram illustrating processes in the form of consecutive activities with responsibilities set out, as well as an indication of applied tools and documents. The use of such a map makes possible the monitoring of the compliance of processes introduced in practice with those mapped, analysing the processes in terms of possible improvements or designing new processes so that their introduction does not disrupt the existing ones. The imperfection of notation is, however, that it does not allow combining processes with objectives (Rosing et al. 2015), which limits the usefulness of maps as tools supporting improvement of management effectiveness.

Obviously, significant knowledge resources concerning an enterprise are possessed by its workers. Well acquainted with the entire organisation and its individual elements, basing on the experience gained and having a very detailed knowledge of individual processes about what may improve the functioning thereof and what will worsen it, workers are sometimes able to convey this knowledge in the way that enables this knowledge to be used in the decision-making process. The tools applied for this purpose in enterprises mainly include mechanisms for gathering improvement suggestions from workers or information regarding non-conformities they notice. Such tools certainly contribute significantly to solving important problems in an enterprise's operation. However, although research confirms their effectiveness (Andriulo and Gnoni 2014; Lawani et al. 2017), each suggestion is considered separately and the knowledge of possible improvements and identified risks is with a significant delay subject to comprehensive analysis in the context of, for example, the OSH management system.

3.1.2 INTRODUCTION TO THE FUZZY COGNITIVE MAPS METHOD

Without questioning the usefulness of the above-mentioned management supporting tools, it is suggested to consider the application of another one facilitating integration of knowledge about the enterprise from all available sources, i.e. fuzzy cognitive maps (FCM) modelling. FCM modelling is based on the concept of a fuzzy set, i.e. a set composed of elements belonging to it to a certain degree, which is determined by what is referred to as the membership function. The membership function is defined as a curve that illustrates the degree of membership of each point of the deliberative space to a given fuzzy set (Olszewski 2006). Fuzzy set A defined in a certain space X is defined as a set

$$A = \left\{ \left(x, \, \mu(x) \right) : x \in X \right\},$$

where $\mu : X \rightarrow (0, 1)$ is a function of membership to the fuzzy set (Dworniczak 2003).

The concept of a fuzzy set may be explained on the example of a set of enterprises the workers of which are involved in improving occupational safety and health. The involvement would be measured as the percentage of workers who submitted at least one OSH improvement proposal to the management during the year. Using the classic logic – bivalent – according to which an object may or may not be a part of a set, it would be appropriate to indicate the percentage of workers submitting proposals that would allow the given enterprise to be included in the set under consideration. Failing to achieve this value would be tantamount to the fact that an enterprise does not belong to the set. For example, if one assumes that in enterprises where workers are involved at least 80% of them submit proposals for improvement within a year, the fact that 79.9% of the workers in the given enterprise make proposals would mean that it does not belong to this set. Although such a distinction allows for the precise classification of each object into one of the sets, it should be noted that a rigid limit value setting is unnatural.

By defining the set of enterprises the workers of which are involved in improving OSH as a fuzzy set, it is possible to approach the assessment of the enterprise in a more flexible manner, in accordance with the rules of fuzzy logic. Figure 3.1 shows the function of membership to a fuzzy set of enterprises the workers of which are involved in improving occupational safety and health. The value of the function $\mu(x)$ for a given percentage of workers submitting proposals for improvement indicates the extent to which the enterprise belongs to the set. If less than 50% of the workers have submitted proposals for improvement during the year, the membership of the set is within the interval (0, 0.4]. If improvement proposals have been made by between 50% and 60% of workers, the enterprise's membership of the set is in the interval (0.4, 0.85]. When the share of submitting workers has accounted to more than 60% but less than 80%, the enterprise's membership of the set is in the interval (0.85, 1). When no worker has made a proposal for improvement, the enterprise belongs to the set is in the degree equal to 0. If at least 80% of the workers have submitted proposals for improvement, the enterprise belongs to the set is in the degree equal to 1 (Skład 2018).

The basis of the FCM method is the belief that reality is often too complicated to be described only by means of numbers. First of all, many elements that constitute this reality are difficult to measure and even if some attempts are made to measure them, the measures are usually applied to one or several dimensions of a given element and do not refer to all of them. Secondly, accurate measurement is often disproportionately time- and cost-intensive, in relation to the value of information that is obtained.

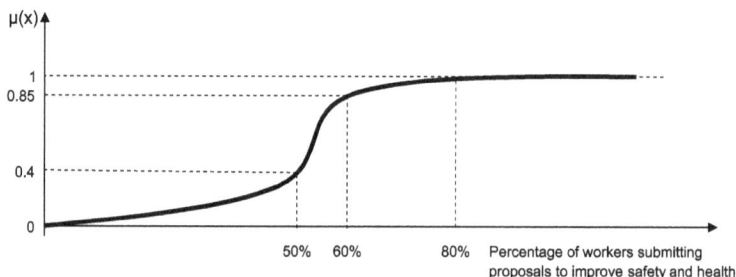

FIGURE 3.1 Membership function of the fuzzy set of enterprises where workers are involved in improving the safety and health conditions. (reprinted, with permission, from: Skład, A. Modelowanie systemów zarządzania bhp z wykorzystaniem metody rozmytych map kognitywnych i wskaźników wiodących – ujęcie teoretyczne. [Modelling of OSH management systems using fuzzy cognitive maps and leading indicators – a theoretical approach]. Bezpieczeństwo Pracy – Nauka i Praktyka nr 2:11–15)

The concept of FCM is based on the assumption that accurate knowledge about particular elements expressed in a specific unit of measure is not always required to make a decision. Verbal and comprehensive specification is quite sufficient to describe and further analyse the examined elements of reality, including the relations between them. The FCM method makes it possible for verbally expressed knowledge to be processed in a way that can support decision making.

The fuzzy cognitive map is a model of a certain fragment of reality and presents knowledge about the cause and effect relationships observed in it. It consists of objects and impacts exerted by these objects on each other. The specificity of models created with the use of fuzzy cognitive maps is that the objects from which these models are composed and the impacts between them have the properties of fuzzy sets. Thus objects can be equivalents of abstract and immeasurable concepts such as events, actions, goals, values (Groumpos 2010; Stylios and Groumpos 1999), characteristics of systems, variables that affect them and their states, as well as system inputs and outputs (Leon et al. 2010). The strengths of impacts between the objects express the degree of causality, i.e. the extent to which the improvement of one object will improve (positively impact) or worsen (negatively impact) the other object.

The fuzzy cognitive map can be presented in two ways: graphically and mathematically. Graphically, the map is a set of objects connected by arrows that symbolise the impacts of individual objects on each other. An example of a fuzzy cognitive map is presented in Figure 3.2, which illustrates the FCM showing the hypothetically considered consequences of introducing a programme dedicated to modifying unsafe behaviours in an enterprise. On the one hand, thanks to the expected increase in knowledge of workers' behaviour, the introduction of the programme increases the effectiveness of undertaken preventive actions and, consequently, makes it possible to reduce the number of accidents and near-misses that negatively affect the efficiency of the enterprise. On the other hand, it is also expected that the tasks under the programme will be carried out by workers, which will increase their workload and, both directly and through increased time pressure resulting in an increase in the number of accidents and near-misses, will result in a decrease in efficiency. In order to express the strengths of impacts exerted between the objects, the symbols "+" and "-" and their multiples were used in Figure 3.2. These symbols have the following meaning: "+", small positive impact; "++", medium positive impact; "+++", large positive impact; "-", small negative impact; "--", medium negative impact; and "---", large negative impact.

A FCM is built and then used to run simulations in order to:

- fully understand the situation;
- forecast its further development; and
- examine alternative scenarios for actions aimed at improving the situation or preventing it from deteriorating.

The fuzzy cognitive maps method was proposed by B. Kosko (1986), who observed that a large part of the knowledge used by people consists of descriptions of the causes of phenomena and various classifications. Thus, in principle, the object of

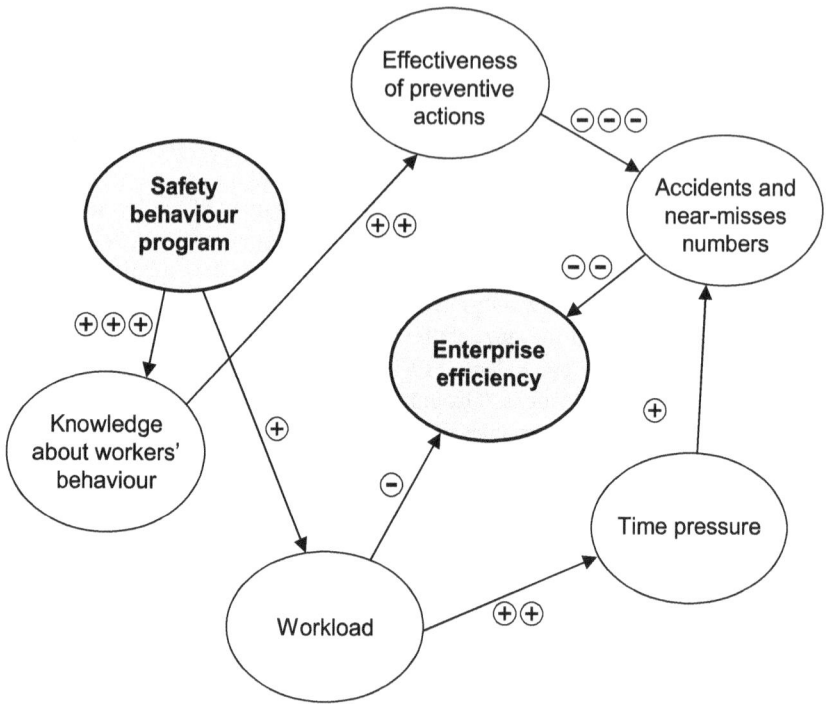

FIGURE 3.2 Example of fuzzy cognitive map

knowledge is characterised by words the full sense and meaning of which can rarely be expressed by numbers. Kosko (1986) noted that the use of verbal descriptions ("fuzzy knowledge representation"), makes it possible to present different points of view and show the complexity of the given phenomenon. On the other hand, it complicates the further processing of such knowledge, e.g. to use it in the decision-making process. Fuzzy cognitive maps are supposed to facilitate processing of descriptive (fuzzy) knowledge, e.g. for decision making.

The fuzzy cognitive map is based on the knowledge of experts who, applying their competences and experience, describe in words the examined fragment of reality, including the assessment of individual elements of the model and the impact thereof on each other. Experts can be both people who have professional theoretical and practical knowledge of a given fragment of reality (doctors, engineers, scientists) and people who know only its practical aspects (e.g. are part of this reality or are observers of it) (Gray et al. 2014).

The use of fuzzy logic allows for the conversion of verbal assessments into numerical values within what is referred to as the defuzzification procedure. Thanks to this, the model using the FCM can be presented in the form of a mathematical notation as a vector of object values and a matrix of values of impacts' strengths exerted by objects. Objects values are in the interval (0, 1) and values of impacts' strengths are in the interval (−1, 1). Presenting the model in this form makes possible conducting simulations with its use. The simulation consists in multiplying the vector of objects values by a matrix of values of impacts' strengths. A single operation is called an iteration, whereby iterations can be performed until the vector of object values stops changing.

To ensure that the objects values during the simulation do not exceed 1, what is known as the activation function, which is most often hyperbolic or sigmoid, is used (Bueno and Salmeron 2009). The result of the simulation is a new vector of object values, which is a forecast of how these values will develop in the future as a result of exerted impacts. A decrease in values means the deterioration of objects and its increase means their improvement. The simulations are conducted in accordance with formula 3.1.

Formula 3.1. New Vector of Object Values

$$
\begin{bmatrix} A_1 & A_2 & \dots & A_n \end{bmatrix}_{new} = f \left(\begin{bmatrix} A_1 & A_2 & \dots & A_n \end{bmatrix}_{old} \times \begin{bmatrix} W_{11} & \dots & W_{1n} \\ W_{21} & \dots & W_{2n} \\ \vdots & & \vdots \\ W_{n1} & \dots & W_{nn} \end{bmatrix} \right)
$$

Source: Adapted from (Pelaez and Bowles 1996).

The method of fuzzy cognitive maps – as proposed by its inventor, Kosko (1986) – was to model systems composed of objects which cannot be precisely defined. According to the author's intention, the method should be used first of all in social sciences to model factors influencing the development of events in specific political, military or social situations, and then to conduct simulations, the results of which would facilitate decision making in solving problems in the modelled situations. However, numerous examples of the application of the fuzzy cognitive maps method available in the literature confirm the universality of the method as a tool that also enables modelling of complex socio-technical and technical systems and conducting simulations, the results of which support decision making in controlling these systems.

3.1.3   REVIEW OF THE FUZZY COGNITIVE MAPS RESEARCH
IN THE FIELD OF MANAGEMENT

The dynamics of changes taking place within the enterprises and in their environment make managing these enterprises more and more difficult, because it requires constant monitoring of these changes and reacting to them. It seems that an FCM-based approach might enjoy increasing interest from the management of enterprises, for whom comprehensive knowledge is essential for making accurate decisions. Previous applications of FCM confirm usefulness thereof in the decision-making processes.

FCMs were used to, inter alia, support decision-making processes in the planning of military actions, aimed at achieving the intended results. The objects in the model were the intended effects (operational goals) and various actions that could directly or indirectly support achieving these goals. The overall objective of the FCM development was to identify alternative courses of action to achieve the operational goals and to choose the best option from among them. The use of the fuzzy cognitive maps method has enabled the simulation of taking individual actions to observe the extent to which they will affect the achieving of operational goals. By analysing the results of the simulations, decision-makers were able to determine what outcomes would be achieved by taking specific actions. In addition, they could anticipate the unplanned effects of their actions, which occurred as a result of numerous complex impacts these action exerted (Yaman and Polat 2009).

The fuzzy cognitive maps method was used to model and assess the factors influencing trust in virtual enterprises.* Basic factors were identified and a methodology was proposed to take into account the dynamic nature of trust in the development of such enterprises (Wei et al. 2008).

The supply chain was modelled using the fuzzy cognitive maps method. Researchers placed 18 objects corresponding to the various types of stocks in the supply chain on the map. They divided them into three categories: external variables that change randomly, e.g.: the inventory consumption, variables depending on management decisions, e.g. inventory safety stock, and internal variables, e.g. stock levels. The values of objects in the chain were estimated based on the data transmitted by RFID technology (i.e. using radio waves).[†] Then, the researchers proposed to conduct two types of analysis. The first type was forward analyses (what-if analysis), which makes it possible to determine the effects on internal variables in the supply chain of a sudden increase or decrease in external and decision making variables. Secondly, they presented the possibility of conducting backward analyses to determine the causes (changes in values of external or decision-making variables) that led to a sudden change in an internal variable. The use of FCM has enabled optimisation of the supply chain operation (Kim et al. 2008).

Another study used the fuzzy cognitive maps method to model reverse logistics[‡] in the supply chain. RFID technology was used to collect real-time inventory data of returns of chilled food containers. The conclusions of the simulations allowed to adjust the parameters of the supply chain and improve its efficiency (Trappey et al. 2010).

In the field of environmental management ecosystems were modelled using the fuzzy cognitive maps method. It was noted that the modelling process involving different stakeholders (residents, authorities, potential investors, scientists, etc.) can be used as a communication and mutual learning tool for stakeholders that supports decision making to facilitate sustainable ecosystem development (van Vilet et al. 2010).

---

* Virtual enterprises – voluntary co-operation between several legally independent enterprises based on trust; all co-operating partners provide their resources, core competencies and knowledge in order to respond more quickly to market needs and act more and more internationally (Wei et al., 2008).
† RFID technology allows the data to be read from an RFID tag on a logistics object (e.g. pallet or product) when it is within reach of the RFID reader. This data can then be saved in the IT system. RFID technology makes possible automatic monitoring of, inter alia, stock levels.
‡ Reverse logistics – planning, implementation and control of the efficient and cost-effective flow of materials, inventory in progress, final goods and related information from the place of consumption to the place of their creation in order to recover value or dispose of them properly [Merkisz-Guranowska 2010).

The fuzzy cognitive maps method was also used in participatory management of urban areas. Based on the experiences of the authorities and residents, factors influencing the urbanisation process were identified, such as the size of the population, the amount of waste generated in the city, infrastructure and health services. These factors were placed as objects on the map, and then simulations were performed in which the values of individual objects were changed and the impact of these changes on the other objects of the model was forecasted. It was proposed that the city authorities should discuss the results of the simulations during a meeting with the inhabitants and, following these results, decide together with the inhabitants on the directions of development of the studied city (Khan and Quaddus 2004).

Another study used the fuzzy cognitive maps method to develop what is known as the proactive Balanced Scorecard.* In order to do that, key performance indicators were identified for each of the four strategic perspectives and then analysed in terms of how they affect each other. In this way a network of Key Performance Indicators (KPIs) was created in the form of a model using FCM. The results of simulations carried out on this model may support the execution of a enterprise's strategy, as they show how a change in the value of one indicator affects the change in the value of other indicators (Chytas et al. 2011).

FCM has also been used in safety management research, including safety and health management. The FCM model was used, inter alia, for FMEA (failure mode and effects analysis) analysis of the high-pressure spray system in a nuclear reactor. The aim of the study was to develop a tool to support decision making in an emergency situation. The authors of the study – on the basis of forecasts developed using the fuzzy cognitive maps method – presented the effects of specific failures and the causes thereof. They proposed developing emergency procedures for nuclear power plant on this basis. In addition, according to their suggestions, the model they developed could be used by operators to run simulations based on information about the current status of system elements. As a result, operators would be able to supervise the proper functioning of the reactor in a more accurate and thoughtful manner (Espinosa-Paredes et al. 2009).

FCM has also been used to investigate the impact of various factors in a food processing plant on human reliability at work. The aim of the study was to identify the most effective and efficient ways to improve occupational safety and health. Thirty-three factors affecting human reliability have been identified, including those relating to working environment, work space, machinery and tools, physical, mental and judgemental load, information and confirmation as well as indication and communication. Then, on the basis of expert assessments, the impacts between the individual factors were determined. Based on the results of a simulation carried out on a model, it was concluded that the two most important factors influencing human reliability are: "loud noise" and "incorrect/unsuitable or lack of indication or communication".

---

* The Balanced Scorecard presents an enterprise's strategy in the form of a set of measurable objectives necessary to achieve the enterprise's mission. It consists of a set of measures distinguished from four perspectives: financial, customer, internal processes and development and their mutual interaction (see Chapter 4).

Identifying these factors helped to decide on the priorities for action to improve safety performance in the enterprise (Bertolini 2007).

The research also used a combination of the AHP method with fuzzy inference. In one of these studies a model was built to assess how much potential for errors that workers may make is related to a given work process. The model covered an enterprise manufacturing car parts. Using FCM, an assessment of the individual processes was carried out and those that should be improved first were identified (Dagdeviren and Yuksel 2008).

It was proposed to use FCM to analyse the relationship between the causes of occupational accidents and types of injuries. The study was conducted in a refinery. Sixteen factors were identified as contributing to accidents at work (lack of attention and concentration, haste, inadequate risk assessment, unsafe behaviour, fatigue, inadequate training, defective equipment, non-compliant premises, lack of control, poor worker well-being, low worker motivation, inadequate procedures, lack of information, inadequate personal protective equipment, poor working conditions, and systemic problems). Twelve types of injuries of varying degrees of severity were also identified. Then, the experts were asked to assess the impacts between all 28 objects. Most of the positive impacts identified by the experts occurred between different causes and injuries. However, some causes were related to other causes (e.g. fatigue). On the basis of the identified impacts and the values of their strengths presented in the form of a matrix, the root causes of the most frequent accidents could then be determined in order to take the necessary preventive measures (Bevilacqua et al. 2012).

The aim of another study was to determine the factors that most impact the resilience of a petrochemical plant. Based on the review of the literature, nine such factors were identified: management commitment, reporting culture, learning culture, awareness, preparedness, flexibility, teamwork, redundancy, and fault-tolerance. On the basis of expert assessments, the impacts between the individual factors were determined using a linguistic variable. Then, using FCM simulations, it was studied the improvement of the enterprise's resilience level achieved by improving individual factors. Calculations were performed which indicated that the preparedness is the factor exerting largest positive impact on the resilience. The second most positive impact on the enterprise's resilience was the awareness, and the last one was the redundancy (Azadeh et al. 2014).

FCM was also used to assess the impact of 27 factors related to occupational safety and health and environmental management on worker productivity, accident rates and job satisfaction. The study was conducted in a gas refinery. The experts were asked to identify the impacts between the different factors and between the factors and productivity indicators, accidents at work rates and job satisfaction index. Based on simulations carried out using FCM, it was found that a lack of documented work instructions is one of the main reasons for low job satisfaction, low productivity and high accident rates. Other reasons include: lack of fresh air, inadequate workstation arrangements, excessive temperature in the workplace, non-ergonomic machines and tools, and lack of instructions and safety training. The factors that are most affected by other factors are stress and time pressure, and these factors, together with high accident rates and low job satisfaction, are the ones that determine low productivity to the greatest extent. The results of the analysis were used to develop a set of leading indicators and it was proposed to use them for proactive productivity management in the enterprise (Asadzadeh et al. 2013).

A similar study was also carried out in an oil and gas transport enterprise. The management system in this enterprise was assessed in terms of the extent to which it meets the requirements of OHSAS 18001:2007, ISO 14001:2004 and ILO-OSH 2001 standards. The system was evaluated according to 13 elements: (1) leadership and commitment; (2) law, regulation and other requirement; (3) capability, training and consciousness; (4) work permit; (5) inner review and audit; (6) risk identification, evaluation and control; (7) equipment integrity; (8) management programme; (9) emergency plans; (10) document control; (11) negotiation and communication; (12) performance measurement and monitoring; and (13) community and public relationship. The following were used as the performance indicators of the integrated environmental and OSH management system: workers' job satisfaction, workers productivity and the enterprise's reputation. Simulations carried out on an environmental and OSH management system model using FCM has shown that among the examined elements of the system, leadership and commitment has the greatest impact on improving system's performance. However, the largest number of impacts is exerted in the system, on its basic elements, such as: work permit (4), emergency plans (9), document control (10), and community and public relationships (13). The most complex network of impacts occurs between these objects and the workers' productivity, which shows how challenging is ensuring productivity in the enterprise (Kang et al. 2016).

This review of research shows that the fuzzy cognitive maps method is a recognised and frequently used tool for management research, including the modelling of safety-relevant systems at the workplace. The application of the fuzzy cognitive maps method makes possible the inclusion in the system models of objects that are equivalent to failures and errors and their causes and consequences, normal or incorrect operation of one of the subsystems or processes of which the system is composed, as well as productivity measures of the examined systems. Models of systems created using the fuzzy cognitive maps method offer the ability to forecast the operation of systems in specific circumstances, which greatly facilitates their design and adaptation to current needs. They also enable the forecasting of short- and long-term consequences of sudden changes in the individual system elements and the planning of emergency procedures if these changes lead to unsafe situations.

Numerous studies conducted using the fuzzy cognitive maps method confirm its usefulness for modelling systems that consist of difficult-to-measure objects, such as: error, incompatibility, event, organisational factors, process, and the subjective perceptions of a group of people. In the case of the studies described above, carrying out these studies using other methods would be very expensive and time-consuming. Based on the fuzzy cognitive maps method, it has been possible to examine the modelled systems in a simple way, based on reliable knowledge and extensive experience of experts, and to present a number of constructive conclusions on the basis of which it is possible to manage these systems and plan their further development.

To sum up, the fuzzy cognitive maps method offers very wide possibilities of modelling systems and forecasting changes in their operation. It is worth using this method in management, especially when the tools used so far do not sufficiently support decision making in an enterprise and do not provide answers to emerging questions.

3.2　DEVELOPMENT OF A MODEL OF THE OSH MANAGEMENT SYSTEM USING THE FUZZY COGNITIVE MAPS APPROACH

3.2.1　BASIC ASSUMPTIONS OF A FCM-BASED MODEL

In this section, the FCM modelling to support decision making in the field of occupational safety and health in enterprises, in particular, to ensure the effectiveness of OSH management systems, has been described. The proposed approach is based on the development of an FCM OSH management system model for a given enterprise and then on conducting analyses and simulations using it. A model of an OSH management system can be developed, depending on the enterprise's needs, in descriptive and graphic form or as a mathematical notation. Presentation of the FCM in mathematical terms requires appropriate collection of "fuzzy" information and making calculations which will enable its presentation with numbers.

The central object in the model of an OSH management system is safety performance. Other objects correspond to processes implemented in the system and to elements of the enterprise context,* i.e. factors exerting negative or positive impact on safety performance and processes.

The development of a model of an OSH management system in descriptive and graphic form consists of several phases. They can be divided into a preparation phase and a modelling phase. The preparation phase includes the selection of a team of experts and the developing of evaluation and assessment scales to be used in the modelling process. In addition, in this phase the goal of the modelling is determined and objects can (at least partially) be identified to be included in the model. The preparation phase is conducted by managers and workers responsible for improving the OSH management system, with the possible participation of specialists in the FCM method.

In turn, the modelling phase is carried out by a team of experts selected during the preparation phase. Taking into account the specific goal of the work and the objects proposed earlier, the team develops a model of the OSH management system: it defines the final set of objects to be used in the model, determines impacts between the objects and assesses the objects and strengths of impacts exerted by the objects.

3.2.2　PREPARATION PHASE

The overall objective of modelling and simulation using the model is to support decision making in the management of OSH within the enterprise, i.e. to improve the

---

* According to ISO 9000 (ISO 2015), understanding the context of an organisation is a process. As a result, factors that impact the sense of the organisation's existence, goals and ability to develop sustainably are identified. The process takes into account internal factors such as values, culture knowledge and performance of the organisation and external factors such as legal, technological, competitive, market, cultural, social, and economic environments.

effectiveness of the OSH management system and/or prevent the possible deterioration of its effectiveness. This can be achieved by introducing specific goals which can be determined taking into account the enterprise's current challenges in the field of OSH management. The basis for the formulation of these goals are the current decision-making problems faced by those responsible for the maintenance and improvement of the system, e.g.:

- determination of priorities for action;
- choosing the best solution from several solutions available; and
- identification of areas where immediate preventive action is needed.

If the adopted modelling goal is precisely formulated, e.g. by defining a set of decision-making options considered by the management, it is worth proposing a set of objects to be included in the model already in the preparation phase. Such objects may be individual decision options and their positive and negative consequences.

The role of experts in the process of modelling OSH management systems can be entrusted to workers who know the system and have broad knowledge about the enterprise. They should be experienced in practical problem solving. External experts may also be invited to co-operate. When choosing people for a team of experts, the planned purpose of their work should be taken into account. If the purpose is to solve a specific problem related to the functioning of the system, it is worth choosing people who understand this issue and, if possible, have experience in the field the problem relates to. Experts should present different points of view and be able to argue their position. On the other hand, they should be able to listen to others, seek compromise and not be guided by the narrowly understood needs of their own organisational unit within the enterprise. A good expert is a person who perceives the problem through the prism of benefits for the enterprise and its stakeholders. Thus, it is worth choosing people trained in soft skills, such as the ability to communicate effectively, negotiate, work as a team, solve conflicts, manage their own time, and motivate themselves and others to work or think creatively.

The manager responsible for the affected area of the decision-making issue may also be a member of the expert team. The manager's participation in modelling is valuable to themselves as it enables them to deepen their knowledge of the area being managed and to confront what they know with the knowledge and experience of people from other organisational units and/or outside the enterprise. In addition, as a participant in the discussion, they will be able to make a more informed final decision based on the recommendations developed by the team.

The team of experts should be between 3 and 10 people. When choosing experts, it is worth taking into account their diversity, both in terms of their place in the organisational hierarchy (ordinary workers and managers) and the type of tasks performed (operational and administrative). In particular, persons directly involved in the maintenance and improvement of an OSH management system (including those employed in the OSH unit, process owners, auditors or the management representative (if appointed), may participate in the work of the team. If the issues that the team resolves are related to human resources management, it is worthwhile for the human resources department to be involved in its work. Issues concerning matters of importance to the whole enterprise should be discussed with the participation of top management. However, issues related to the functioning of individual departments should be discussed with the support of their management and/or workers. The involvement of persons from operational units is of particular importance in the case of teams carrying out work in exposure to agents which are strenuous, harmful or particularly dangerous, to ensure that the recommendations developed by the team do not inadvertently increase exposure to these agents. As regards persons from outside the enterprise, these may include representatives of the labour inspectorate or local authorities (Bevilacqua et al. 2012), as well as consultants in the field of OSH or the enterprise management in general.

In order to facilitate communication and a consistent description of the objects identified and their relationships, FCM proposes to use scales, both to assess objects and to evaluate strengths of their impacts. The scales provided to the experts should consist of several simple linguistic values (e.g. small, medium, and large) and give an opportunity to express practical knowledge about modelled phenomena and relationships without using numbers.

The scales should be extensive enough to ensure that a distinction can be made between different assessments of objects and assessments of the strengths of their impacts. The scale for assessing the strength of impact should include values expressing both positive and negative impacts.

Examples of scales are presented here:

a) for the assessment of objects corresponding to management processes, a five-stage scale composed of values
   - definitely compliant with OSH management system's requirements,
   - compliant with OSH management system's requirements,
   - hard to say,
   - non-compliant with OSH management system's requirements,
   - definitely non-compliant with OSH management system's requirements;
b) for the assessment of object corresponding to safety performance, a five-stage scale composed of values
   - very high,
   - high,
   - medium,
   - low,
   - very low;
c) for the assessment of objects corresponding to context factors exerting an impact on processes and safety performance, a five-stage scale composed of values
   - very low intensity,
   - low intensity,
   - medium intensity,
   - high intensity,
   - very high intensity,
   according to which the intensity of these factors should be determined;

d) for the assessment of the impacts exerted by the objects on each other, a 13-stage scale composed of values
- impact close to 1,
- very strong positive impact,
- strong positive impact,
- average positive impact,
- low positive impact,
- very low positive impact,
- no impact,
- very low negative impact,
- low negative impact,
- average negative impact,
- strong negative impact,
- very strong negative impact,
- impact close to −1,
  according to which it should be determined how strong the impacts exerted by particular management processes and factors are.

3.2.3 Modelling Phase

The experts should understand the principles of FCM modelling as well as be acquainted with the scales developed for use in the modelling process. They should be explained the goal for which the team has been set up and in what form they are to present the results of their work. Before starting the modelling process, a deadline for completion of the work and its schedule is also defined, taking into account the availability of individual team members. The group should indicate the person responsible for moderating the discussion. His or her task is to ensure that everyone can speak and express his/her position. Moreover, the moderator of the discussion should be able to establish and enforce the rules of the discussion.

The first task of the experts is to select the objects that the system model will consist of. Extraction of objects should be closely related to the purpose of the team's work, i.e. the decision problem under consideration, its sources, and areas where its consequences are visible. The model should cover the whole system, but the degree of detail of the individual elements may vary, depending on the specific needs of the enterprise with regard to the expected results of modelling and forecasting using the model.

The central element of the model is an object corresponding to safety performance. It is a reference point for assessing the effectiveness of the system*, understood as the ability of the system to improve safety performance. Moreover, among the objects, elements of the system should be distinguished, i.e. processes that

---

* According to ISO 45001 (ISO 2018), effectiveness is the degree to which planned activities are implemented and planned results achieved, while the intended result of an OSH management system is to increase safety performance, i.e. continuous improvement of OSH performance, meeting legal and other requirements, and achieving OSH objectives. Therefore, the measure of effectiveness of an OSH management system is the change in safety performance as a result of system operation.

constitute it. A process is defined as a set of inter-related or interacting actions that transform inputs into outputs.

The model should also include objects corresponding to elements forming the enterprise context, i.e., according to ISO 45001 (ISO 2018), "external and internal issues that are relevant to its purpose and that affect its ability to achieve the intended outcomes of its OSH management system". External context factors may include the business environment (e.g. cultural, social, legal, financial, technological, competitive, etc.), the emergence of new contractors, new technologies, new regulations and new professions and relations with external stakeholders. In turn, internal context factors may be, inter alia, related to the introduction of new technologies, changes in the organisational structure, resources such as capital, time, human resources, processes, systems, and technologies. Lack of knowledge about the existence, areas of impacts and strengths of impacts of these factors prevents the objectives of an OSH management system from being achieved, making it ineffective, as the factors may interfere with the planned course of individual management processes in the system in an uncontrolled and unpredictable way and contribute to the deterioration of safety performance.

When defining the set of objects to be included in the model, the experts take into account the objects proposed in the preparation phase, but may, at their own discretion and perception of the OSH management system, group those into larger or divide into smaller objects. It should be remembered that the more detailed the model of the system (many precisely defined objects), the more work-intensive further phases of the model development is. On the other hand, the more extended the model is the more detailed forecasting results it can deliver.

The team of experts analyses each object one by one, putting it in pairs with all the other objects. By exchanging their knowledge and experience and using examples from practice, the experts successively decide whether the examined object affects the other object in a pair and, if they agree that there is an impact, they discuss the value of the impact's strength using the scale defined in the preparation phase. The experts analyse n^2-n pairs of objects, systematically recording the results of their findings.

In addition to assessments of the values of impacts' strengths exerted among objects in the system model, the experts also express their assessments of individual objects. The criteria for this assessment vary, depending on the type of the object. As far as objects corresponding to processes are concerned, their assessment is conducted on the basis of criteria defined for these processes in the standard underpinning the OSH management system implemented in the enterprise (e.g. ISO 45001). Knowing and understanding these criteria, the experts discuss what value, according to the appropriate scale defined in the preparation phase, to assign to the individual objects. In their assessment of an object corresponding to safety performance, experts consider in particular indicators relating to accidents at work, near-misses and unsafe behaviour and the exposure of workers to strenuous, harmful and hazardous agents as well as psychosocial factors in the working environment. The criteria for assessing objects corresponding to the elements forming the enterprise context depend on the nature of these elements. When assessing the context, experts should therefore be guided primarily by their feelings about the intensity of the individual elements and relate them to past experience and knowledge of what the minimum and maximum intensity may be.

Experts' assessments of objects and impacts are subjective and this is characteristic and purposeful in the case of the fuzzy cognitive maps method, which is designed to collect, organise and process that part of the experts' knowledge that cannot be used to manage a enterprise using other methods. According to the FCM concept, this knowledge is imprecise but comprehensive and can be gathered in a relatively fast and cheap way (Gray et al. 2014; Pelaez and Bowles 1996).

The result of modelling is a graph showing the individual objects of the model and their impacts exerted on each other expressed by arrows. If there are so many elements and their impacts that the graph becomes illegible, a simplified version of the graph can be prepared, e.g. by eliminating the arrows symbolising impacts with relatively lowest values. Another way to present the modelling results is to put them in the table, where in the first column and in the first line all objects included in the model are listed in sequence, and in remaining cells the values of impacts' strengths are written, e.g. in the cell in the third column and in the second line the value of the strength of impact exerted by object No 1 on object No 2 is determined, and in the fifth column in the sixth line the value of the strength of impact exerted by object No 5 on object No 4 is determined.

An alternative to expert teamwork is the identification of objects and impacts and their assessment by each individual expert separately. However, the adopting of this approach requires that at some point in time "partial" models are collected and aggregated. It is also worth remembering that working in a group allows the experts to share their knowledge, views and experiences, which may foster more thoughtful assessments, while individually developed expert assessments will reflect unilateral views, built on their subjective feelings.

3.2.4 Mathematical Notation of the Model

One of the basic principles of modelling by experts using FCM is the use of linguistic values. Avoiding numerical values is intended to ensure that experts do not try to "force" quantify something they cannot count and/or measure, and that they do not use fragmented, familiar measures to assess complicated and complex objects. Linguistic scales are therefore available to experts.

However, this does not mean that it is not possible to convert verbal expert assessments into numbers. Transforming the model into a mathematical notation is possible using the principles of fuzzy logic, and, above all, what are known as linguistic variables. Linguistic variables are variables the values of which are not numbers but words or sentences in natural language. Each language value has a label. The label is a word or sentence belonging to a set of linguistic terms, and its meaning is a fuzzy subset. The linguistic variable is used to approximate the characteristics of phenomena that are too complex or imprecisely defined to be unambiguously described using conventional quantitative terms (Herrera and Herrera-Viedma 2000).

A linguistic variable is also defined as a four (N; V; U; I), where N is the name of the variable (e.g. workers' competence in the field of occupational safety and health), V a set of linguistic values (e.g. high, medium, low), U a deliberative space, what is known as *universe* (e.g. from 0 to 100% of correct answers during health and safety examinations), and I a set of interpretation of terms – a set of membership functions (Diering et al. 2014). Figure 3.3 shows an example set of membership functions of a linguistic variable describing workers' competence in the field of occupational safety and health.

The linguistic variable shown in Figure 3.3 is defined by three membership functions: function $\mu 1$, representing the degree of membership to a set of workers possessing low OSH competence; function $\mu 2$, representing the degree of membership to a set of workers possessing medium OSH competence; and function $\mu 3$, representing the degree of membership to a set of workers possessing high OSH competence. On the basis of the linguistic variable defined in Figure 3.3, it can be concluded, for example, that the competences of a worker who gave 40% of the correct answers during the test belong in 0.5 degree to the set of low competences and in 0.5 degree to the set of medium competences.

There are several methods to sharpen fuzzy values (*low, medium, high*), i.e. to determine their numerical values, based on the membership functions. For sharpening (defuzzification) linguistic variables, the following methods are used: the weighted mean of maximum method (Bowles and Pealez 1995) and the centre of area method (Papageorgiou et al. 2009).

There are many examples of linguistic variables used for FCM modelling in literature. These variables usually contain from five to nine values, and in the case of the variables used to assess values of impacts' strengths, from 10 to 18 (with the same number of positive and negative values (Papageorgiou et al. 2009; Stylios and Groumpos (2004); Stylios et al. 2008; Bevilacqua et al. 2012; Georgopoulos et al. 2003; Dağdeviren et al. 2008).

In order to transform the descriptive and graphic model of the system developed by the expert team, using linguistic variables, the linguistic values of objects and

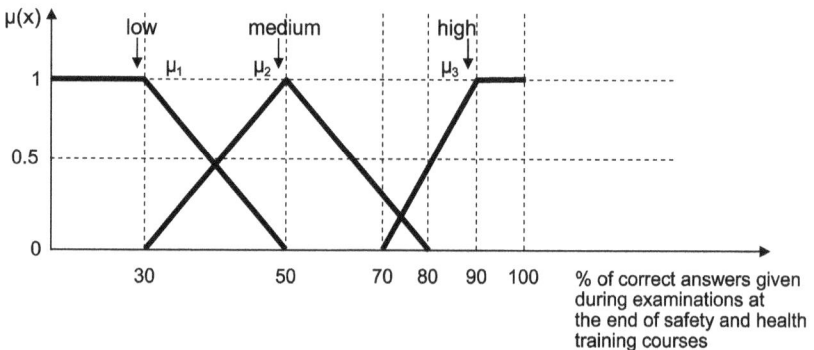

FIGURE 3.3 A set of membership functions of a linguistic variable describing workers' competence in the field of OSH. (Reprinted, with permission, from: Skład, A. 2017. Modelowanie i prognozowanie wpływu poprawy procesów zarządczych w systemie zarządzania bhp na poziom bezpieczeństwa w przedsiębiorstwie. [Modelling and forecasting the impact of improvement of management processes in the OSH management system on the safety performance of the enterprise]. PhD diss., CIOP-PIB)

impacts' strengths established by the experts should be successively subjected to the process of defuzzification and replaced with numerical values. Objects are numbered C_1, C_2...C_n, and their values form a vector, ordered by the order of object numbering $C = (c_1, c_2,...c_n)$.

Values of impacts' strengths transformed into numbers are written in a matrix with the dimensions n x n, where n = number of objects, in such a way that in the first line the values of impacts' strengths exerted on all objects from C1 to Cn by the first C1 object are written in sequence, in the second line – the values of impacts' strengths exerted on all objects from C1 to Cn by the second C2 object in the model are written successively, and in the last line – the values of impacts' strengths exerted on all objects from C1 to Cn by the last Cn object in the model are written successively, with the value of the impact's strength exerted by a given object on itself always equal to 0.

If the enterprise has decided to create system models by individual experts separately, partial models are aggregated to develop the final model. If the partial models consist of a different number of different objects, and thus the sizes of the vectors of the object values and the matrices of values of the impacts' strengths are different, then they must be "added up". For this purpose, it is first necessary to develop a set of objects containing the objects distinguished in all models. Then the values of those objects that have been indicated by more than one expert are averaged. As far as objects indicated in single partial models are concerned, values specified in these models are assigned to them. In the next step, an aggregated matrix of values of impacts' strengths is developed, including rows and columns corresponding to the impacts exerted by and on the individual objects of the aggregated model, respectively (Khan and Quaddus 2004; Stach et al. 2010). If all the experts are equally reliable in the opinion of the management, the simplest method of summing up matrices of the same dimensions is to calculate the arithmetic mean of the expert assessments of the value of the impacts' strengths occurring between individual objects. If there are grounds to differentiate the reliability of individual experts involved in system modelling, appropriate weights can be assigned to their assessments with values in the interval (0, 1), and then, in order to obtain the final value of the impact strength between the two objects, a weighted average of the assessments received from the experts is calculated (Georgopoulos et al. 2003; Stach et al. 2010; Stylios and Groumpos 1999; Stylios et al. 1997). Where a given expert is a valued authority in the functioning of certain objects of the system, only some of their estimated values of impacts' strengths may also be given correspondingly higher weights (Stylios and Groumpos 1999).

In the literature, a matrix of values of impacts strengths exerted in an OSH management system between 15 processes and these processes and safety performance is available. This matrix, developed by nine experts dealing with the practical and scientific system approach to OSH management,* can be used by extending it via adding objects corresponding to elements of the context in a given enterprise, which can facilitate and accelerate the process of system modelling in this enterprise (Skład 2019a).

---

* The values of the impact forces determined verbally by the individual experts have been defuzzified and averaged.

Figure 3.4 presents a model of a system developed for a specific enterprise using the above-mentioned matrix. The model consists of 26 objects: 15 of them correspond to management processes in an OSH management system, 10 correspond to factors that exert negative impacts on the operation of these processes, and one corresponds to safety performance. Objects are marked with symbols

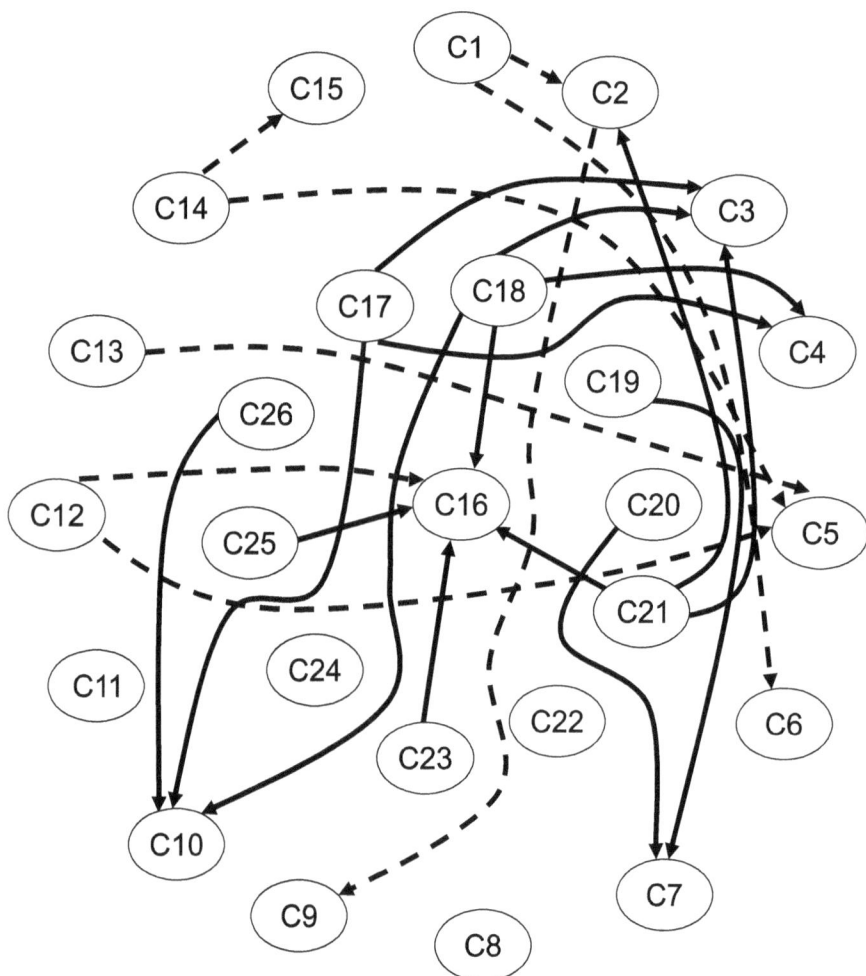

FIGURE 3.4   A model of an enterprise OSH management system using the fuzzy cognitive maps method. (Reprinted, with permission, from: Skład, A. 2019b. Model systemu zarządzania bezpieczeństwem i higieną pracy w podejściu procesowym. Zastosowanie metody kognitywnych. Część III. Przykłady rozmytych map poznawczych obrazujących zależności pomiędzy wyjściami z procesów zarządczych w systemie zarządzania bezpieczeństwem i higieną pracy. [A model of an occupational health and safety management system in a process approach. Application of cognitive mapping method. Part III. Examples of fuzzy cognitive maps illustrating relationships between management processes in an occupational health and safety management system]. CIOP-PIB)

C1 to C 26 and these are: C1, Leadership; C2, OSH policy; C3, Organisational roles, responsibilities and authorities; C4, Consultation and participation of workers; C5, Management review; C6, Communication and information; C7, Training and competence management; C8, Taking advantage of OSH opportunities; C9, Planning and achieving OSH objectives; C10, Occupational risk management; C11, Emergency preparedness and response; C12, Incident, non-conformity and corrective actions; C13, Internal audit; C14, Monitoring and performance evaluation; C15, Documented information; C16, Safety performance; C17, Lack of workers due to retirements; C18, Insufficient number of workers caused by difficulties in attracting and retaining new workers; C19, Time constraints for investment/modernisation projects; C20, Investment projects conducted outside the enterprise's premises; C21, Implementation of new regulations/policies; C22, Changes in production processes; C23, Replacement of machinery and equipment; C24, Equipment and machine failure; C25, Repairs; and C26, Diverse safety culture level of subcontractors. Due to the large number of impacts between individual objects identified in the model, only impacts with relatively highest and lowest values of impacts' strengths (exceeding 0.6 or below −0.6) were included in the map, in order to keep the figure legible. Positive influences are marked with a dotted line and negative ones with a continuous one.

3.3 APPLICATION OF THE MODEL TO SUPPORT DECISION MAKING

3.3.1 Choice of the Application Method

The essence of modelling by the team of experts is to use the developed model to achieve the assumed goals. Depending on the model's notation (descriptive and graphic or mathematical), it can be used in various ways to support decision-making processes in an OSH management system. Applying on the basis of a descriptive and graphic model requires relatively tedious analysis of individual objects in terms of their influence, but it can be successfully carried out by the same team of experts and does not require specialist knowledge of fuzzy logic.

Converting a descriptive and graphic model into a mathematical notation enables simulations to be performed using it, but requires the use of linguistic variables and the ability to defuzzify fuzzy linguistic values into numbers. Furthermore tools to perform matrix calculations (multiplication of vector by matrix) are required. In most cases, it is necessary to have the support of external consultants who will join the expert team and provide the necessary tooling facilities. However, the defuzzification and calculation tool allows for very fast simulation and analysis of many different decision options in a short time.

Taking this into account, it seems that descriptive and graphical modelling is recommended for small enterprises with simple management systems and no significant resources for tools to improve them. On the other hand, complex systems in large enterprises should rather be analysed on the basis of mathematical notation models and using appropriate calculation tools.

3.3.2 Inference Based on a Descriptive and Graphic Model

In order to prepare recommendations on the basis of a descriptive and graphic model, the first step is to carry out a model analysis involving, in particular:

- identification of objects corresponding to the factors exerting the strongest negative and positive impact on the object corresponding to safety performance;
- identification of objects corresponding to the processes assessed as non-compliant;
- indication of objects corresponding to the processes on which the largest number of objects exert negative impacts with relatively highest strengths' values;
- identification of objects exerting the strongest negative and positive impact on objects corresponding to the processes;
- identification of objects exerting the greatest number of negative and positive impacts on objects corresponding to the processes;
- analysing the possibility of weakening the objects corresponding to the factors exerting a negative impacts on the operation of processes and strengthening the objects corresponding to the factors exerting a positive impacts; and
- considering different scenarios for improving individual management processes in the system in terms of how their improvement will affect other processes and safety performance.

In the next step, using the accumulated and structured knowledge about the safety performance, the processes of the OSH management system and the factors that exert impact on the processes and safety performance, it is necessary to discuss:

- the priorities for improving management processes in the OSH management system;
- possible solutions to improve safety performance and processes' operation; and
- possible solutions to the problem the solution of which was the goal of the expert team.

The discussed solutions should be placed as additional objects (sets of objects) in the model and, when analysing their impact on all existing objects, the pros and cons of these solutions should be considered. The team's work should result in recommendations for improving the effectiveness of the OSH management system, in particular with regard to the problem defined as the work goal. The recommendations should primarily concern the improvement of management processes in the system and the implementation of measures to improve these processes, together with an indication of the anticipated effects of the proposed changes. The team of experts can recommend several alternative solutions to the problem. Each solution should be described along with its advantages and disadvantages.

3.3.3 THE APPLICATION OF THE MODEL IN MATHEMATICAL NOTATION FOR SIMULATIONS

The application of models in mathematical notation makes possible conducting simulations, which result in forecasting future values of objects. As described previously, the simulation consists of multiplying the vector of objects values by a matrix of values of impacts' strengths. The result is a new vector of object values. If the new (forecasted) value of the object is lower than the current one, it means that the corresponding process, factor or safety performance will deteriorate. If it the forecasted value is higher, it will improve.

Therefore, in the model of an OSH management system, conducting such simulations enables one to obtain a forecast of improvement or deterioration of safety performance and individual processes in the system. In addition, by comparing the forecast values with the current values, it is possible to determine for which objects the forecast change is the greatest, which becomes particularly important in the case of a decrease in value. The forecasting of a significant fall in the value of an object corresponding to safety performance should be an important warning signal for the management of the enterprise and prompt them to take immediate preventive action. Similarly, the forecasting of a significant reduction in the value of one or more objects corresponding to processes means that, without radical corrective measures, the enterprise's OSH management system will soon fail to meet the requirements of the introduced standard.

As in the case of the descriptive and graphical model, the model in mathematical notation is also used to improve the effectiveness of the OSH management system, with special attention to the problem the team of experts has been asked to solve. Assuming that the effectiveness of an OSH management system is its ability to improve the safety performance in an enterprise, and that the measure of effectiveness of an OSH management system is a change in the value of the object corresponding to safety performance, it is necessary to consider the possibility of taking actions that will result in an increase in the value of this object and to analyse their impact on the whole system. A specific way of planning improvement and prevention measures using FCM is to develop what are known as "what-if" scenarios. The development of these scenarios consists of analysing the model in terms of identifying possible ways of preventing the situation from deteriorating, e.g. by decreasing the value of those objects that have the strongest negative impact on objects corresponding to safety performance or processes, or increasing the value of those that have a positive impact. These scenarios may also include the inclusion of new objects in the models, equivalent to the actions the implementation of which may improve the effectiveness of the given OSH management system.

While conducting simulations, a model configuration (objects of which it consists and the values thereof) is sought, at which the highest value of an object corresponding to safety performance and the highest values of objects corresponding to processes are obtained. Establishing this configuration is the basis for the formulation of recommendations by a team of experts who, when comparing the current model with the target configuration:

- indicate factors that have a negative impact on the system, which should be eliminated in the first place;

- identify factors with a positive impact on the system that should be strengthened;
- recommend which processes in the system should be improved as a matter of priority in order to ensure the greatest effectiveness of the system; and
- recommend other measures to improve the effectiveness of the system.

As in the case of model analysis in descriptive and graphical form, also simulating with the application of a model in mathematical notation, the experts can present several alternative solutions to the decision problem under consideration.

The result of the work of a team of experts modelling an OSH management system is a set of recommendations, including their pros and cons, relating to the goal of this work. However, the final management decision is taken by a person with appropriate authority, i.e. a specific management representative. This decision does not have to be in line with the team's recommendations, for example because the costs required to introduce the recommendation may turn out to be too high. This does not mean, however, that the team's work has been performed unnecessarily or will not be used. On the contrary, the recommendations developed by the team are, together with the rationale therefore, a source of valuable information about the pros and cons of all the decision options under consideration, and by choosing an option other than the recommended one, the manager is aware of its possible disadvantages and/or negative consequences and can take action to reduce or eliminate them.

3.4 SUMMARY

The FCM method makes possible the collection and processing of valuable information about the OSH management system resulting from the knowledge and experience of the enterprise's workers. Modelling using this method may be carried out regularly at specific time intervals. The model and forecasts developed using it are an excellent starting point for the management review process. Modelling can also be used on an ad hoc basis to solve current decision-making problems. Depending on current needs, only the part of the model that is relevant to the problem may be used for forecasting or the model as a whole. Modelling can also be used in worker participation processes: to collect opinions and identify potential problems. Once developed, the model of the OSH management system can be updated and improved. It should be emphasised that this requires relatively small amount of work and financial resources as compared to the implementation of, for example, a set of indicators for the assessment of the OSH management system.

Modelling using FCM can provide important support for the implementation of the OSH management system compliant with the requirements of ISO 45001, as far as section 4 on the enterprise context is concerned. The flexibility of this modelling method, in terms of diversity of the objects to be considered, enables the entire context to be analysed comprehensively and its content to be updated quickly by adding new elements, subtracting those that have ceased to influence the OSH management system, and changing the intensity of the elements and the values of impacts' strengths they exert.

It appears that a particularly valuable feature of modelling using FCM is the ability to process and disclose tacit knowledge available in the enterprise. Although research shows that having tacit knowledge in an organisation can bring benefits, especially in a situation of considerable competitive pressure (Lecuona and Reitzig 2014); on the other hand, difficulties in transferring this knowledge are pointed out (Seidler-de Alwis and Hartmann 2008). Of course, every workers meeting is an opportunity to exchange tacit knowledge and contributes to its consolidation among the workers, but the principles of modelling, including above all the use of scales to assess the value of objects and the impacts occurring between such, force the structuring of this knowledge and a specific way of recording it. This can greatly facilitate the use of the potential of tacit knowledge in the enterprise.

In addition, the modelling process strengthens communication between workers and contributes to a better mutual understanding of the problems faced by workers of individual departments, as well as to their knowledge of the enterprise as a whole.

Despite many successful applications of the fuzzy cognitive maps method for system modelling, it is also important to remember about some difficulties related to its application, the most important of which is the lack of unambiguous methodology for conducting simulations using mathematical notation of the model (Espinosa-Paredes et al. 2009) which applies in particular to the lack of guidelines for the selection of the threshold function. Enterprises interested in using the FCM may also encounter difficulties in providing appropriate tooling facilities for simulation. However as the accumulation, consolidation and application of diverse expertise through this method can significantly improve the effectiveness of OSH management systems, it is suggested to apply it in business practice and take advantage of the opportunities it offers.

REFERENCES

Andriulo, S., and M. G. Gnoni. 2014. Measuring the effectiveness of a near-miss management system: An application in an automotive firm supplier. *Reliability Engineering and System Safety* 132:154–162. https://reader.elsevier.com/reader/sd/pii/S095183201400 177X?token=20A5D57F598CE93FB13133CF495FF3CA52E856EC739EEC2B4F2FD6 2EE36FB64D3BC2FB756F8ADA7D1576592E3A6C71A2 (accessed June 17, 2019).

Asadzadeh, S. M., A. Azadeh, A. Negahban, and A. Sotoudeh. 2013. Assessment and improvement of integrated HSE and macroergonomics factors by fuzzy cognitive maps: The case of a large gas refinery. *Journal of Loss Prevention in the Process Industries* 26(6):1015–1026. https://reader.elsevier.com/reader/sd/pii/S095042301300 082X?token=7F50A6759168E5065717A5D54BE02B2D16BF0FF38C193B65DD2A85 FAFF976DEF3056A669200DE373FBCF8F461D55B718 (accessed July 23, 2019).

Azadeh, A., V. Salehi, M. Arvan, and M. Dolatkhah. 2014. Assessment of resilience engineering factors in high-risk environments by fuzzy cognitive maps: A petrochemical plant. *Safety Science* 68:99–107. https://reader.elsevier.com/reader/sd/pii/S092575351400 071X?tokcn=0D39A75E6D9518F34B5604BD7E23B5CE23B42B597F7B33F498B1CB 305FCE0980B142E6C1CBA325939B65064040AA8AEB (accessed July 23, 2019).

Bertolini, M. 2007. Assessment of human reliability factors: A fuzzy cognitive maps approach. *International Journal of Industrial Ergonomics* 37(5):405–413. https://reader.elsevier.com/reader/sd/pii/S0169814107000170?token=D148A91E590398A0D85 6C2564E2BAE175DB72667A100D9DB50A2B65638008765BD77C3CCDA6958E 67ECF6F8E098D9A21 (accessed July 24, 2019).

Bevilacqua, M., F. E. Ciarapica, and G. Mazzuto. 2012. Analysis of injury events with fuzzy cognitive maps. *Journal of Loss Prevention in the Process Industries* 25(4):677–685. https://reader.elsevier.com/reader/sd/pii/S0950423012000265?token=5213D7C24 819743B64FA874524F5AA95C449D99F2A0FCBB2A5F9A1C77798D8BFF7908 116DCB762AFCA8F5FBA1A29650D (accessed August 1, 2019).

Bowles, J. B., and C. E. Peldez. 1995. Fuzzy logic prioritization of failures in a system failure mode, effects and criticality analysis. *Reliability Engineering and System Safety* 50(2):203–213. https://reader.elsevier.com/reader/sd/pii/095183209500068D?token= 4E9CC2B01F62D0A79BF4DEF90BB727CF42D891CE9C22781DFE42F9CE7B49 1F598AF51E751A1B6143FF34AC403820D726 (accessed June 17, 2019).

Bueno, S., and J. L. Salmeron. 2009. Benchmarking main activation functions in fuzzy cognitive maps2. *Expert Systems with Applications* 36(3):5221–5229. http://isiarticles.com/bundles/Article/pre/pdf/1309.pdf (accessed July 26, 2019).

Chytas, P., M. Glykas, and G. Valiris. 2011. A proactive balanced scorecard. *International Journal of Information Management* 31(5):460–468. http://www.mihantarjomeh.com/wp-content/uploads/2014/03/A-proactive-balanced-scorecard.pdf (accessed June 10, 2019).

Dağdeviren, M., and I. Yuksel. 2008. Developing a fuzzy analytic hierarchy process (AHP) model for behavior-based safety management. *Information Sciences* 178(6):1717–1733. https://reader.elsevier.com/reader/sd/pii/S0020025507005130?token=5F61619E4 E91BEADBBBC80F75E5EF0E361CAB01E52482AE28D10AF248A00503602573 CA92026AB99E9518A7E71F15025 (accessed August 9, 2019).

Dağdeviren, M., I. Yűksel, and M. Kurt. 2008. A fuzzy analytic network process (ANP) model to identify faulty behavior risk (FBR) in work system. *Safety Science* 46(5):771–783. https://reader.elsevier.com/reader/sd/pii/S0925753507000367?token=8551C3E66 27E87F0F45B432223F52B1ABEA38BA9BA13D45EC8CAF0DB50AB68E09225E 628543F3A5965510180014B17A7 (accessed September 16, 2019).

Diering, M., A. Kujawińska, K. Dyczkowski, and M. Rogalewicz. 2014. Logika rozmyta w ocenie alternatywnych systemów pomiarowych jako jeden z kierunków rozwoju MSA. (Fuzzy logic in the evaluation of alternative measurement systems as one of the directions of MSA development). *Innowacje w zarządzaniu i inżynierii produkcji* T. 2:348–359.

Dworniczak, P. Zbiory rozmyte dla początkujących. Zapis odczytu wygłoszonego na XXX Szkole Matematyki Poglądowej Osobliwości, w styczniu 2003 roku. (Fuzzy sets for beginners. Record of the reading given at the XXX School of Mathematics of Perceptual Peculiarity, in January 2003). https://smp.uph.edu.pl/msn/31/dworn.pdf (accessed July 25, 2019).

Espinosa-Paredes, G., A. Nunez-Carrera, A. Vazquez-Rodriguez, and E.-G. Espinosa-Martinez. 2009. Modelling of the High Pressure Core Spray Systems with fuzzy cognitive maps for operational transient analysis in nuclear power reactors. *Progress in Nuclear Energy* 51(3):434–442.

Georgopoulos, V. C., G. A. Malandraki, and C. D. Stylios. 2003. A fuzzy cognitive map approach to differential diagnosis of specific language impairment. *Artificial Intelligence in Medicine* 29(3):261–278. https://reader.elsevier.com/reader/sd/pii/S09 33365702000763?token=854F8B7ECF7078D808F088D16D76036DEB4EB8CA4A0D 6791284F81AB2A48D66EF2C87510F381C7840118AA042E9F9C75 (accessed August 9, 2019).

Gray, S. A., E. Zanre, and S. R. J. Gray. 2014. Fuzzy cognitive maps as representations of mental models and group beliefs. In: *Fuzzy Cognitive Maps for Applied Sciences and Engineering: Intelligent Systems Reference Library 54*, ed. E. I. Papageorgiou, 29–48. Springer-Verlag, Berlin Heidelberg. https://link.springer.com/content/pdf/10.1007 %2F978-3-642-39739-4.pdf (accessed July 26, 2019).

Groumpos, P. P. 2010. Fuzzy cognitive maps: Basic theories and their application to complex systems. In: *Studies in Fuzziness and Soft Computing, Volume 247: Fuzzy Cognitive Maps*, ed. M. Glykas, 1–22. Springer-Verlag, Berlin Heidelberg. https://link.springe r.com/content/pdf/10.1007%2F978-3-642-03220-2.pdf (accessed June 3, 2019).

Herrera, F., and E. Herrera-Viedma. 2000. Linguistic decision analysis: Steps for solving decision problems under linguistic information. *Fuzzy Sets and Systems* 115(1):67–82. https://reader.elsevier.com/reader/sd/pii/S016501149900024X?token=61B688FD5 FC69FFABD101FD2D9543C5C06C33A227873590EFFE66CABC343CDE7296AE AD7BFD5B947E56DE1D8F991422Ev (accessed July 31, 2019).

Husain, S., Y. Ahmad, M. Sharma, and S. Ali. 2017. Comparative analysis of defuzzification approaches from an aspect of real life problem. *IOSR Journal of Computer Engineering* 19(6):19–25. Ver. III. http://www.iosrjournals.org/iosr-jce/papers/Vol19-issue6/Ver sion-3/D1906031925.pdf (accessed July 26, 2019).

ISO (International Organization for Standardization). 2015. ISO 9000:2015, Quality management systems – fundamentals and vocabulary. International Organization for Standardization, Geneva, Switzerland.

ISO (International Organization for Standardization). 2018. ISO 45001:2018, Occupational health and safety management systems: Requirements with guidance for use. International Organization for Standardization, Geneva, Switzerland.

Kang, J., J. Zhang, and J. Gao. 2016. Improving performance evaluation of health, safety and environment management system by combining fuzzy cognitive maps and relative degree analysis. *Safety Science* 87:92–100. https://reader.elsevier.com/reader/ sd/pii/S0925753516300297?token=FB80849FBF62654FE5E627BC05704241D5 7D99A055FC10D53476CEED9D9CBDA00AD374706208041E6622D464E769F93D (accessed June 28, 2019).

Khan, M. S., and M. Quaddus. 2004. Group decision support using fuzzy cognitive maps for causal reasoning. *Group Decision and Negotiation* 13(5):463–480. https://link.sp ringer.com/content/pdf/10.1023%2FB%3AGRUP.0000045748.89201.f3.pdf (accessed July 23, 2019).

Kim, M.-C., C. O. Kim, S. R. Hong, and I.-H. Kwon. 2008. Forward-backward analysis of RFID-enabled supply chain using fuzzy cognitive map and genetic algorithm. *Expert Systems with Applications* 35(3):1166–1176. https://reader.elsevier.com/reader/ sd/pii/S0957417407003429?token=2409A1C0ECEDF45D9A6BF1199A2E56F47F AD6C462EB8B6040F276DAFBCF5C1E1B817B725E50B6B039F94DB5F3389B457 (accessed September 5, 2019).

Kosko, B. 1986. Fuzzy cognitive maps. *International Journal of Man-Machine Studies* 24(1):65–75.

Lawani, K., B. Hare, and I. Cameron. 2017. Developing a worker engagement maturity model for improving occupational safety and health (OSH) in construction. *Journal of Construction Project Management and Innovation* 7(2):2116 – 2126. https://journal s.co.za/docserver/fulltext/jcpmi_v7_n2_a11.pdf?expires=1568621830&id=id&acc name=guest&checksum=FC7FA44A7ACC698CE33594BDBCB3EFFC (accessed September 10. 2019).

Lecuona, J. R., and M. Reitzig. 2014. Knowledge worth having in "excess": The value of tacit and firm-specific human resource slack. *Strategic Management Journal* 35(7):954–973. https://onlinelibrary.wiley.com/doi/epdf/10.1002/smj.2143 (accessed July 23, 2019).

León, M., C. Rodriguez, M. M. García, R. Bello, and K. Vanhoof. 2010. Fuzzy cognitive maps for modelling complex systems. In: *Advances in Artificial Intelligence. 9th Mexican International Conference on Artificial Intelligence. Proceedings Part I*, ed. G. Sidorov et al. 166–174, Springer-Verlag, Berlin Heidelberg. https://link.springer.com/content/p df/10.1007%2F978-3-642-16761-4.pdf (accessed September 16, 2019).

Merkisz-Guranowska, A. 2010. Logistyka recyklingu odpadów jako jeden z elementów systemu logistycznego Polski. (Waste recycling logistics as one of the elements of the Polish logistics system). *Prace Naukowe Politechniki Warszawskiej* z 75:89–96.

Olszewski, R. 2006. Systemy wnioskowania rozmytego (FIS) jako narzędzie nieliniowej generalizacji numerycznego modelu terenu. (Fuzzy Inference Systems (FIS) as a tool for non-linear generalization of the numerical terrain model) *Polski Przegląd Kartograficzny* Tom 38, nr 4:316–325. http://ppk.net.pl/artykuly/2006402.pdf (accessed June 10, 2019).

Papageorgiou, E. I., A. Markinos, and T. Gemptos. 2009. Application of fuzzy cognitive maps for cotton yield management in precision farming. *Expert Systems with Applications* 36(10):12399–12413. https://reader.elsevier.com/reader/sd/pii/S0957417409003960?token=3EE3385275B6DD66D5BA0FDB51D1B7B144DC142751B8FBB45B8245AC78E72A23E47BA9FE5CB252C6F3DB722AE1004DB4 (accessed August 9, 2019).

Pelaez, C. E., and J. B. Bowles. 1996. Using fuzzy cognitive maps as a system model for failure modes and effects analysis. *Information Sciences* 88(1–4):177–199. https://reader.elsevier.com/reader/sd/pii/0020025595001611?token=7BB39C10BD050B8884B58A772C0418F375B128CD8DADE9D1967E011EACDE5CB2C8F9AF93877CE142867D39BF50BF005A (accessed July 23, 2019).

von Rosing, M., S. White, F. Cummins, and H. de Man. 2015. Business process model and notation – BPMN. In: *The Complete Business Process Handbook: Body of Knowledge from Process Modelling to BPM. Vol. 1*, 429–453. Morgan Kaufmann. https://www.omg.org/news/whitepapers/Business_Process_Model_and_Notation.pdf (accessed July 3, 2019).

Seidler-de Alwis, R., and E. Hartmann. 2008. The use of tacit knowledge within innovative companies: Knowledge management in innovative enterprises. *Journal of Knowledge Management* 12(1):133–147. https://ai.wu.ac.at/~kaiser/literatur/Wissensmanagement-2/the-use-of-tacit-knowledge.pdf (accessed Sptember 16. 2019).

Skład, A. 2017 Modelowanie i prognozowanie wpływu poprawy procesów zarządczych w systemie zarządzania BHP na poziom bezpieczeństwa w przedsiębiorstwie. (Modelling and forecasting the impact of improvement of management processes in the OSH management system on the safety performance of the enterprise). PhD diss., CIOP-PIB.

Skład, A. 2018. Modelowanie systemów zarządzania BHP z wykorzystaniem metody rozmytych map kognitywnych i wskaźników wiodących – Ujęcie teoretyczne. (Modelling of OSH management systems using fuzzy cognitive maps and leading indicators – a theoretical approach). *Bezpieczeństwo Pracy – Nauka i Praktyka* 2:11–15.

Skład, A. 2019a. Assessing the impact of processes on the occupational safety and health management system's effectiveness using the fuzzy cognitive maps approach. *Safety Science* 117(August):71–80. https://reader.elsevier.com/reader/sd/pii/S0925753518319428?token=31C0B88C52E4C1EA7235E7C85561115E650AF158A9EE0820EC7822C474CDF5BEF9E9E7DD6BCF16C7889E90C04F8D8ACA (accessed September 16, 2019).

Skład, A. 2019b. Model systemu zarządzania bezpieczeństwem i higieną pracy w podejściu procesowym. Zastosowanie metody map kognitywnych. Część III. Przykłady rozmytych map poznawczych obrazujących zależności pomiędzy wyjściami z procesów zarządczych w systemie zarządzania bezpieczeństwem i higieną pracy. (A model of an occupational safety and health management system in a process approach. Application of cognitive mapping method. Part III. Examples of fuzzy cognitive maps illustrating relationships between management processes in an occupational health and safety management system). CIOP-PIB.

Stach, W., Ł. Kurgan, and W. Pedrycz. 2010. Expert-based and computational methods for developing fuzzy cognitive maps. In: *Studies in Fuzziness and Soft Computing Volume 247: Fuzzy Cognitive Maps*, ed. M. Glykas, 23–41. Springer-Verlag, Berlin Heidelberg. https://link.springer.com/content/pdf/10.1007%2F978-3-642-03220-2.pdf (accessed July 23, 2019).

Stylios, C. D., V. C. Georgopoulos, and P. P. Groumpos. 1997. Introducing the theory of fuzzy cognitive maps in distributed systems. In: *Proceedings 12th IEEE International Symposium on Intelligent Control*, ed. K. Ciliz, and Y. Istefanopulos, 55–60, Bogaziqi University, Istanbul, Turkey.

Stylios, C. D., and P. P. Groumpos. 1999. Fuzzy cognitive maps: A model for intelligent supervisory control systems. *Computers in Industry* 39(3):229–238. https://reader.elsevier.com/reader/sd/pii/S0166361598001390?token=D01B263A263A60541D0097618CA9D6A19716B2607FD3E1E0287CBBFC9122FC1C678EDCCD75DB42C3A7BB920F420227E4 (accessed June 11, 2019).

Stylios, C. D., and P. P. Groumpos. 2004. Modelling complex systems using fuzzy cognitive maps. *IEEE Transactions on Systems, Man, and Cybernetics – Part A: Systems and Humans* 34(1):155–162.

Stylios, C. D., V. C. Georgopoulos, G. A. Malandraki, and S. Chouliara. 2008. Fuzzy cognitive map architectures for medical decision support systems. *Applied Soft Computing* 8(3):1243–1251. https://reader.elsevier.com/reader/sd/pii/S1568494607001226?token=CAA07F1A45CBED8985B5568CD2660334F12B1B2654BCAC414D2B3F23E9346129C67FA1CCED3399174ECAF8CE5FCF7801 (accessed June 3, 2019).

Trappey, A. J. C., C. V. Trappey, and C.-R. Wub. 2010. Genetic algorithm dynamic performance evaluation for RFID reverse logistic management. *Expert Systems with Applications* 37(11):7329–7335. https://reader.elsevier.com/reader/sd/pii/S0957417410002976?token=3637D1D342DEACB875F527695FB40F4586B7EA133A099F97E3137F470E03280331192E4F0998AE7E1AE0AD2ED22E9A00 (accessed June 11, 2019).

van Vliet, M., K. Kok, and T. Veldkamp. 2010. Linking stakeholders and modellers in scenario studies: The use of fuzzy cognitive maps as a communication and learning tool. *Futures* 42(1):1–14. https://reader.elsevier.com/reader/sd/pii/S0016328709001360?token=B16A2C3BACB1937C06B0745AF8D5960E46AA3E84A240AD2F9AA3C5AE9B9FAB6B5EDA529D67CEC5ED89E8DC5F5A92EC87 (accessed September 5, 2019).

Wei, Z., L. Lu, and Z. Yanchun. 2008. Using fuzzy cognitive time maps for modelling and evaluating trust dynamics in the virtual enterprises. *Expert Systems with Applications* 35(4):1583–1592. https://reader.elsevier.com/reader/sd/pii/S095741740700382X?token=385F341BF39EEA6B16E640172A14BFFCA3CBE9014C11F8DD475F261CD8C36BC39258EB10B086DAF689652336ADCAA32A (accessed July 26, 2019).

Yaman, D., and. S. Polat. 2009. A fuzzy cognitive map approach for effect-based operations: An illustrative case. *Information Sciences* 179(4):382–403. https://reader.elsevier.com/reader/sd/pii/S0020025508004246?token=AD4B83874A91516D67DA7183FE2F68491DA55FE58F9C2E9B68F580986E2DE0A7B86B1C738BB5C240EAC1E783EB82D6DD (accessed July 23, 2019).

4 Strategic Thinking in OSH Management

Małgorzata Pęciłło

CONTENTS

4.1 APPLICATION OF BALANCED SCORECARD TO OSH MANAGEMENT

4.1.1 THE IDEA OF BALANCED SCORECARD

The methods and tools developed over the years for measuring and monitoring organisational processes are not very effective from the point of view of supporting the performance of organisation if they function as separate entities. In order to make full use of them, it is necessary to create an integrated, holistic system for evaluating the performance of management processes, which would make it possible to link the assessment of their performance to the given enterprise's strategy and resources, and which would enable irregularities to be identified and preventive measures to be taken a priori before problems related to these irregularities arise. A good system for measuring the effectiveness of activities carried out in the organisation should take into account the specificity of each organisational process, its complexity, the multiplicity of factors influencing its introduction, the variety of measures necessary for its measurement and the links between them (Kueng 2000).

The need to link the evaluation of processes to the strategic and operational level is highlighted by Rummler and Brache (1995): they stress the necessity of setting the objectives for each level of management, i.e. organisation, processes and departments and workstations, starting with setting them at the organisation level and ending at the workstation level. Then, measures should be assigned to these objectives. The measures should be appropriate to the specifics of each level of management, with the objectives of the jobs being linked to the objectives of the departments to which they are assigned. Both objectives of the jobs and of the departments should be resulting from the objectives of the processes.

Another concept for an organisation's measurement system was developed in the early 1990s: the Balanced Scorecard (BSC), which shows the cause-and-effect relationships between enterprise's strategy and four distinct perspectives: financial, customer, internal processes and development, thus combining financial and non-financial measures into a comprehensive evaluation system (Kaplan and Norton 1992). The assessment of an organisation using this concept consists of defining the global strategy of the enterprise, and then determining appropriate objectives for each perspective, starting from the highest hierarchical perspective, directly resulting from the strategy, and assigning to these objectives adequate measures, both quantitative and qualitative.

A similar method to the Balanced Scorecard was developed in France as early as the 1930s: the Tableau du Bord de Gestion (TBG) method, which focused mainly on non-financial measures and complemented the data provided by management accounting (Bourguignon et al. 2004). The basic difference between the two methods lies in the fact that the measures in the BSC are presented in a sequence of logically related objectives and indicators that are coherent and complementary (Kaplan and Norton 1996), which is missing in the French method. Another tool that preceded the development of the Balanced Scorecard was the Japanese hoshin kanri method, which has been used since the 1960s (Witcher and Chau 2007).

As it can be observed from the examples quoted, the very idea on which the Balanced Scorecard concept is based is much older; the concept is also evolving (Brown 2006; Tuomela 2000; Olve et al. 2002, Olve et al. 2003; Friedag and Schmidt 2004; Neely at al. 2002). For example, Friedag and Schmidt (2004) extended the Balanced Scorecard by adding six new perspectives linked to enterprise strategy, i.e. the communication, organisational, supplier, shareholder, public and implementation (e.g. software) perspectives. In the view of Friedag and Schmidt, each enterprise should choose the most appropriate combination of proposed perspectives and build its own Balanced Scorecard (BSC) thereon. In turn, the creators of the Skandia Navigator propose relying on five perspectives representing different elements of capital, i.e. development, customers, processes and human, and financial focus (Edvinsson and Malone 1997). Neely et al. (2002) consider that this concept should be extended to include other stakeholders and not focus only on customers. In contrast to Norton and Kaplan's Balanced Scorecard, where the starting point is the enterprise's strategy, Performance Prism, building an evaluation system starts with the identification of stakeholders and their expectations and contributions to the organisation, and only then is defining the strategy to meet the needs and expectations of the stakeholders established (Neely et al. 2002).

The approach proposed by Kueng (2000) is also to expand the circle of process stakeholders and to go beyond the circle limited to customers. In this approach, the basis for determining the objectives of process implementation, and thus the measures of these processes, are the general objectives of the organisation, the competition, and the process stakeholders (i.e. customers, workers and society). The extension of the organisation's evaluation card to include more stakeholders is directly part of the Corporate Social Responsibility (CSR) concept, which is reflected in work on CSR (Panayiotou et al. 2007; Debnath et al. 2018).

The common feature of all the above approaches, despite some differences, is linking measurement systems to selected perspectives (among which the perspective of internal processes is always present, regardless of the type of card) to ensure a balanced and holistic view on the given organisation.

These evaluation cards do not exhaust the topic, but they show that the Balance Scorecard is not so much a tool as a concept to support the introduction of a strategy in an organisation by defining the cause-and-effect relationships between different indicators and perspectives. These models are not ready-to-go solutions, but are only the basis for enterprises to develop their own comprehensive evaluation systems. Information technology development can support the Balanced Scorecard design and business process management. There is a clear tendency to move from flexible methods towards tools and programmes that can be used automatically.

4.1.2　The Balanced Scorecard in OSH

Petersen (2000) pointed out that perhaps the biggest safety problem was, and still is measurement. He indicates that, in hindsight, it can be said that specialists dealing with safety choose ineffective, inadequate and inappropriate measures. There is a need for a system for evaluating occupational safety and health activities which will provide information on both the effectiveness and efficiency of OSH management. It is also intended to make possible, inter alia, rational decision making with regard to occupational safety and health (OSH) and assessments of its effects, comparing the activities currently taken in the area of OSH with those taken in the past and with the activities of other enterprises, as well as comparing the effects of these activities and comparing the current OSH activities with those planned for the future (Toellner 2001). Hence the unflagging interest in measurement methods and tools that can be used in the field of OSH management, including interest in the possibilities offered by the Balanced Scorecard concept (Ingalls 1999; Petersen 2001; Langhoff 2002; Mearns 2003).

The Health and Safety Executive (Mearns et al. 2003) proposes including in its model four perspectives closely related to the classical BSC concept, specifically: the financial perspective, i.e. expenditures on occupational safety and health at work; the perspective of internal customers, i.e. the level of safety culture; the perspective of internal processes, i.e. internal occupational safety and health procedures and instructions; and the perspective of development, i.e. the knowledge and skills in the field of occupational safety and health at work available to the organisation. In turn, the model proposed by Ingalls (1999) is based on four slightly different perspectives: the OSH management system and safety culture, individual worker attitudes and behaviour, the introduction of programmes aimed at improving occupational safety and health, and worker knowledge and skills. Azour et al. (2017) propose a strategy to integrate health, safety, and environment indicators to the new Balanced Scorecard by explaining the cause-and-effect relationship between its six perspectives, adding occupational safety and health and environmental perspectives.

The presented Balanced Scorecard, based on Norton and Kaplan's classic Balanced Scorecard concept (Kaplan and Norton 1996), enables one to look at the implementation of occupational safety and health measures from four perspectives (Figure 4.1):

- the growth and learning perspective, i.e. the key features and skills of workers and procedures applied in the field of occupational safety and health management;
- the perspective of internal processes, i.e. the way in which activities in the field of occupational safety and health are introduced;

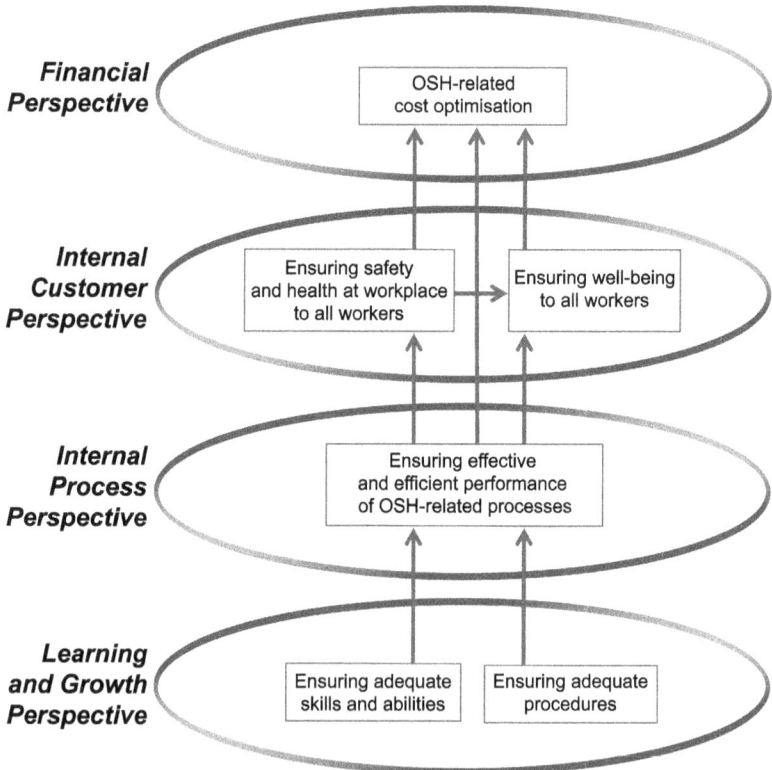

FIGURE 4.1 OSH-related objectives in the Balanced Scorecard (reprinted, with permission, from: Pęciłło 2008. Assessing effectiveness of occupational safety and health management. in Weiss, Godlewska and Bitkowska [eds]. *Nowe trendy i wyzwania w zarządzaniu. Koncepcje zarządzania.* Warsaw: Vizja Press & IT. p. 436)

- the perspective of internal customers, i.e. how workers perceive safety and what level of safety is provided to them; and
- the financial perspective, i.e. expenditures on the introduction of measures in the field of occupational safety and health at work and losses incurred as a result of inappropriate working conditions.

Each of the four perspectives should be assigned its own specific objectives, starting with the financial perspective and ending with the development perspective, and appropriate measures to assess the achievement thereof. Achieving objectives at a lower level should contribute to achieving objectives at a higher level.

The development perspective includes the desired and specialised qualities and skills of the members of the organisation, which ensure the effective performance of OSH management processes; whereby qualities are understood here as those attitudes and behaviours of workers which can be influenced by the organisation and which can shaped as, for example, loyalty, willingness to improve qualifications, level of motivation, willingness to take risks, and individual safety culture. This perspective covers occupational safety and health services and other persons involved in the introduction and execution of OSH management processes in an enterprise, i.e. members of the top management, managers of organisational units, and regular workers and their representatives.

The perspective of internal processes includes the basic processes of OSH management, which should be carried out in each enterprise, i.e.:

- hazard identification and risk assessment process;
- the process of monitoring working conditions;
- the process of taking preventive actions in the field of occupational safety and health (the process of corrective and/or preventive actions);
- the process of training in occupational safety and health; and
- the process of transferring information on occupational safety and health between the participants of the organisation (internal communication process in the field of occupational safety and health).

The efficiency of the processes mentioned above may be measured by their effectiveness, i.e. the degree of achievement of the assumed objectives and the efficiency, which consists of two measures: the time and the cost of the introduction of the process.

The perspective of internal customers applies to all workers of the enterprise. This perspective has two interrelated objectives:

- ensuring safe and hygienic working conditions for all workers of the enterprise, which is measured by, inter alia, indicators such as accidents at work, occupational diseases, non-injury incidents and the number of people working in hazardous conditions; and
- ensuring the well-being of workers, i.e. their well-being, which contributes, among other things, to the productivity and quality of their work.

The financial perspective refers to the costs incurred by the enterprise resulting from improper working conditions and expenditures on the introduction of OSH management processes and the meeting of legal requirements in this respect and economic benefits resulting from these processes.

The main costs related to inappropriate working conditions include the costs of accident insurance, the costs arising from accidents at work and occupational diseases and the costs resulting from work under harmful and strenuous conditions (e.g. the costs of increasing sickness absence and benefits for work under harmful and arduous conditions and the costs resulting from reduced quality and productivity of work).

The primary objective of the financial perspective is to reduce the costs related to occupational safety and health while ensuring that the enterprise will meet the requirements of the applicable legislation in this area.

The Balanced Scorecard looks different if we view safety and health from the perspective of Corporate Social Responsibility. The aim of the financial perspective should be to achieve a high score on the Dow Jones Sustainability Index (DJSI), which provides information about the financial performances of firms which are leaders in their industry on the field of CSR (DJSI Index Family) rather than to optimise costs. Since, as mentioned previously, the Balanced Scorecard objectives are set starting from the highest perspective, i.e. the financial perspective, it is these objectives that determine the final shape of the scorecard and the selection of individual indicators (Figure 4.2).

In the context of CSR, in addition to the processes directly aimed at ensuring occupational safety and health the Balanced Scorecard for the field of OSH should take into account the processes which, in a broader context, treat the well-being of workers and ensure job satisfaction, which is derived from human resources processes, including, in particular:

- recruitment of workers;
- professional adaptation of workers;
- human resources development;
- payroll management;
- motivating of workers;
- enabling workers to participate in the management of the organisation;
- conflict resolution and ensuring good interpersonal relations; and
- the departure of workers.

A natural consequence of the introduction of new processes may be an extension of requirements at the level of the growth and learning perspective.

4.1.3 THE BALANCED SCORECARD AND OSH MANAGEMENT SYSTEMS

The models of an OSH management system may vary depending on normative documents or guidelines, but the general principles and basic elements remain the same (cf. PKN 2004; BSI 2004; BSI 2007; ILO 2001; AS/NZS 1997/2001). These models

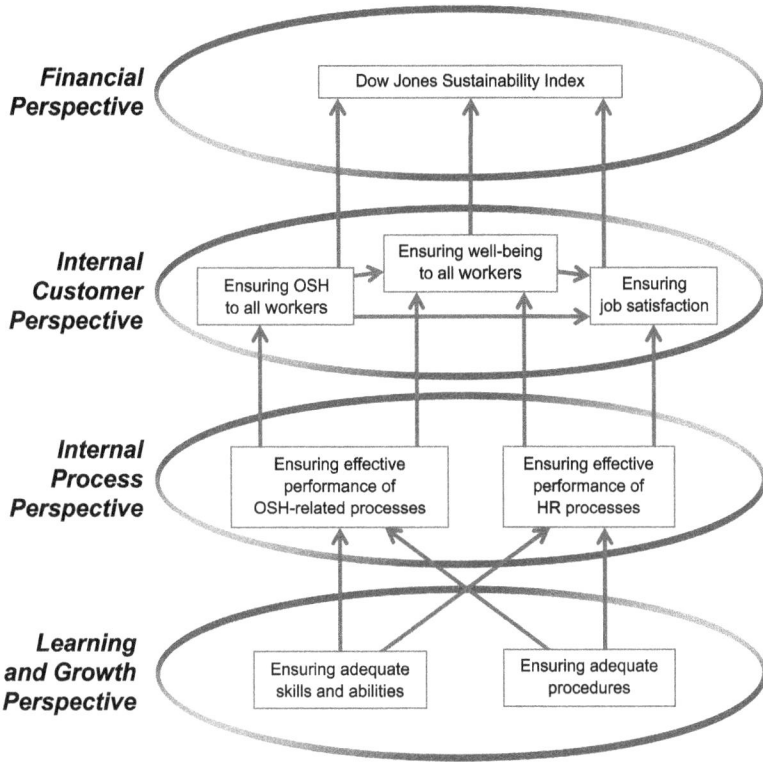

FIGURE 4.2 OSH-related objectives in the Balanced Scorecard (BSC) from the CSR perspective

are based on the concept of continuous improvement. The following elements can be distinguished in them:

- occupational safety and health policy, the involvement of senior management and worker participation;
- planning, including setting general and specific objectives and identifying and meeting legal requirements;
- implementation and operation, including, inter alia, ensuring resources, competence, communication, documentation, co-operation with subcontractors, purchasing and occupational risk management;
- evaluation, which consists of monitoring and measurement, investigation of accidents at work, auditing and review by management; and
- improvement measures, in particular preventive and corrective action and continuous improvement.

A good OSH management assessment system should also provide an answer to the question of whether the type and performance of OSH management processes in an enterprise result from an established OSH policy and whether these processes contribute to the execution of that policy. The assessment system should also make it possible to assess the quality of the resources needed to carry out the individual activities that form part of the OSH management processes.

The identification of links between strategic OSH strategy, general objectives (at the process level) and operational ones (at the workplace level) using a Balanced Scorecard is consistent with the system approach to OSH management. It is recommended that a traditional OSH management system should set safety and health policy (i.e. OSH strategy), then goals for OSH management (i.e. general objectives) and finally targets for the level of jobs and departments (i.e. operational objectives), and then assign appropriate measures to these and define plans for achieving these objectives (Figure 4.3). Such an approach has been applied, among others, in the standards on OSH management systems adopted by the Polish Committee for Standardisation PN-N-18001 (PKN 2001) and PN-N-18004 (PKN 2004). What distinguishes the presented approach is the lack of clear separation of organisational processes and relying on the traditional organisational structure and vertical approach to management as opposed to the process approach and horizontal approach.

The process approach to occupational safety and health management systems has been directly introduced in the ISO 45001 (1SO 2018) standard mentioned above. Introducing this approach requires establishing criteria for process assessment (Figure 4.4).

In addition to processes, this standard, as a first for occupational safety and health management systems, introduces the necessity to analyse the context of the organisation and to identify risks and opportunities for the safety and health management system and for the safety and health of workers and other interested parties, such as subcontractors and visitors.

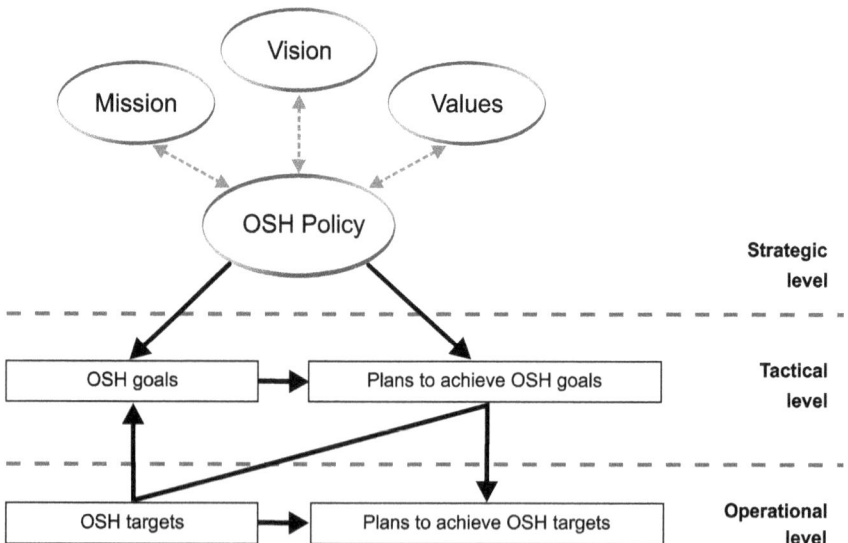

FIGURE 4.3 Objectives in OSH management systems

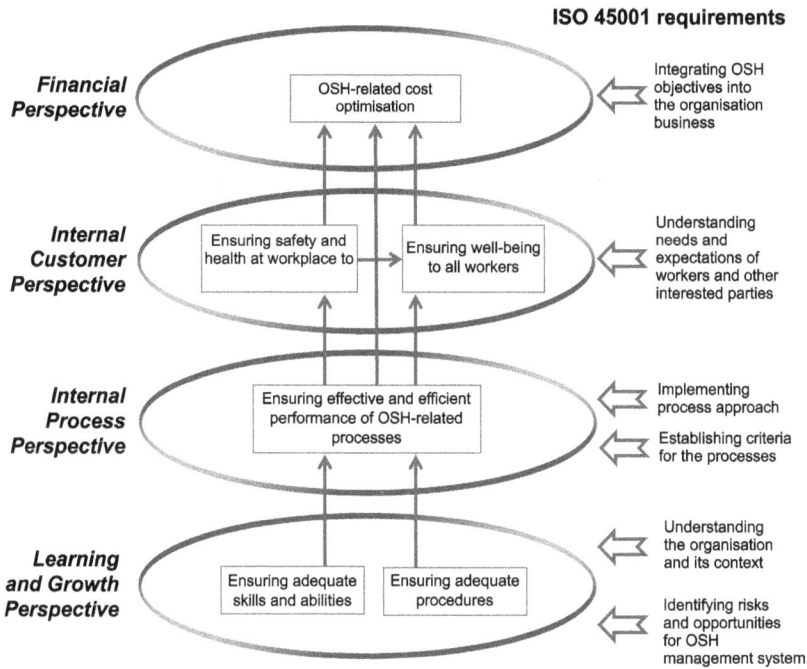

FIGURE 4.4   Requirements of ISO 45001 standard in the context of the BSC

The context of an organisation is nothing more than its external and internal environment, defined as a set of factors and processes that affect the organisation directly and indirectly. The external environment of the organisation consists of (Griffin 2011):

- the macro-environment, including external factors with a broader scope, which similarly affect the activities of all economic entities present on a given market and are usually independent from the organisation, such as geopolitics, internal policies of countries, legislation, the labour market situation, new technologies and the pace of change in technology, available natural resources, energy costs, and climate factors;
- the micro-environment, including external factors which directly affect the enterprise and which may also be influenced to some extent by the enterprise. This environment includes, inter alia, customers, suppliers, local communities, competitors, inspections (e.g. labour inspectors), etc.

The internal environment is created by factors shaped by owners and shareholders, management and workers, and includes resources in the broad sense (human, material, financial, informational, etc.), as well as the procedures applied and organisational culture.

The analysis and understanding of the organisational context are the starting point for the identification of risks and opportunities for an OSH management system and understanding needs and expectations of workers and other interested parties. Identification of risks and opportunities for an OSH management system enables one to ensure adequate resources and proper design of the OSH management system at the level of the learning and growth perspective, which in turn translates into proper performance of OSH-related processes. However, the identification of the needs and expectations of workers is necessary to determine the objectives at the level of internal customers, especially in terms of ensuring their well-being.

Satisfying the ISO 45001 standard requirements on ensuring the integration of an OSH management system into the organisation's overall business should start from the strategic level, i.e. the financial one.

4.1.4　THE PRACTICAL USE OF THE BALANCED SCORECARD – AN EXAMPLE FROM BEHAVIOURAL-BASED PROGRAMMES

Unsafe behaviours have been studied since the beginning of the 20th century, which resulted in initiatives being introduced by enterprises to promote safe behaviours and eliminate unsafe behaviours among workers, guests and subcontractors (Heinrich 1941). An unsafe behaviour modification programme is defined as a set of techniques aimed at encouraging or discouraging workers to/from certain predefined behaviours to prevent accidents at work and occupational illnesses. This means that every activity properly structured and co-ordinated with other activities and aimed at modelling a high safety culture may be actually considered a part of an unsafe behaviour modification programme. Therefore, systems of penalties and rewards, training programmes or information activities aimed at increasing worker awareness of threats and safe performance of work are all elements of unsafe behaviour modification programmes and have been used by enterprises for many years. Behavioural-Based Safety programmes (BBS) have a very specific structure and are usually based on behavioural observation as their central element. The programme itself varies from enterprise to enterprise but there are also common elements of such a programme: the first step of its implementation is training aimed at explaining its legitimacy and course, as well as the issues related to safety culture, and thus improve workers' perception of unsafe behaviours of their own and their colleagues. The next step is to observe and record unsafe and safe behaviours. The observation and registration can be carried out by both regular workers and supervisors, as well as an established team. The observation of unsafe behaviours may take the form of what are known as behavioural audits. At this stage, there should be a modification of unsafe behaviour, i.e. stopping work and instructing the given worker to encourage him/her to change his/her behaviour. Recorded information on unsafe behaviours is analysed in order to identify the causes thereof and propose corrective actions.

For such a designed programme, as well as for any other accident prevention programmes, the BSC can be used as a method of measuring its effectiveness and assessing its suitability for the introduction of the given enterprise's strategy in the field of occupational safety and health (Table 4.1). Some of the measures proposed in this card, which are assigned to particular perspectives, as in other cards found in literature, are difficult to estimate. Information about their level may be obtained from audits, observations, interviews and worker surveys. Such measures include, inter alia, those related to the development perspective.

TABLE 4.1

OSH-Related Aims and Indicators of a BBS Programme in the Balanced Scorecard

No	Perspective	Aim	Indicator
1.	Financial perspective	OSH-related cost optimisation	Costs resulting from improper working conditions
			Costs resulting from investment in OSH
2.	Internal customer perspective	Ensuring safety and health at workplace to all workers	Indicator of frequency of occupational accidents
			Indicator of severity of occupational accidents
			Indicator of non-injury incidents
			Indicator of workers employed in harmful conditions
			Indicator of work-related diseases
		Ensuring well-being to all workers	Level of absenteeism
			Level of presenteeism of sick workers
			Work ability
3.	Internal processes perspective	Ensuring effective and efficient performance of OSH-related processes, e.g.:	Indicators depend on type of given process, e.g.:
		Reducing number of examples of unsafe behaviour	Total number of observations (i.e. safe and unsafe behaviours) per worker observing
		Increasing number of examples of safe behaviours (Unsafe behaviour modification programmes)	Number of examples of safe behaviour per number of examples of unsafe behaviour
			Number of examples of unsafe behaviour per total number of being observed
			Number of corrective measures resulting from observations per number of examples of unsafe behaviour
4.	Growth and learning perspective	Ensuring adequate skills and abilities of workers involved in process performance	Communication skills
			Expertise
			Motivation
			Awareness of OSH related risks
			Knowledge of OSH-related risks
			Safety Culture
			Worker involvement
			Length of service

4.2   SELECTED ASPECT OF RESILIENCE ENGINEERING

4.2.1   Resilience Engineering as a New Concept of OSH Management

The resilience engineering concept was developed in the beginning of the 21st century as a result of a need to find solutions with regard to safety and health management as traditional approach and tools for occupational safety and health seemed to come to the limits of effectively inducing significant increases of safety level by raising expenditures and effort (Shannon and Mayer and Haines 1997). In addition, the idea of resilience was adapted to safety and health management as an answer to modern management theory and practice, strongly focused on efficiency, productivity and effectiveness, as these three parameters turned out to be insufficient in case of disruptions (Woods 2006b). Resilience engineering strongly corresponds with Hollnagel's idea of Safety-II defined as ensuring that things go right and seen as opposed to the traditional approach based on avoiding things that may go wrong (Hollnagel 2014). It is important to keep in mind that what resilience engineering provides is a new perspective for analysing safety issues, but it does not deny traditional methods and measures. It "differs more in the perspective it provides on safety than in the methods and practical approaches that are used to address real-life problems" (Hollnagel et al. 2008).

Resilience engineering defines safety as "the ability to succeed under varying conditions" (Hollnagel 2011). Being obvious and necessary, changes are not seen as a threat, but a natural consequence of activity in evolving environments resulting in positive and negative outcomes. And what allows an organisation to overcome negative outcomes is resilience, which may be defined as the "ability of a system or an organisation to react and recover from disturbances at an early stage, with minimal effects on dynamic stability" (Hollnagel 2006).

According to resilience engineering theory, there are four abilities needed for a system to be resilient: responding ("knowing what to do"), monitoring ("knowing what to look for"), anticipating ("knowing what to expect"), and learning ("knowing what has happened") (Woods 2006a). The abilities are called the "four cornerstones". With some variations resulting from specific traits and different activities, an appropriate level of each cornerstone is necessary and a deficiency in one of them cannot be compensated by a high level of the remaining cornerstones.

4.2.2   Resilience Engineering in OSH Management Systems

A resilient OSH management system is a system that focuses on both failure and success investigation at every stage of management: learning, monitoring, anticipating and finally reacting. In what are referred to as traditional occupational safety and health management systems there are two types of monitoring: reactive and active. Reactive monitoring, according to the resilience engineering concept, is nothing less than learning from past events. In the ILO Guidelines on occupational safety and health management systems reactive monitoring is defined as "checks that failures in the hazard and risk prevention and protection control measures, and the OSH management system, as demonstrated by the occurrence of injuries, ill health, diseases and incidents, are identified and acted upon" (ILO 2001). Therefore, it does not meet the requirements of this concept, as it includes, according to the traditional approach to the subject, investigating adverse events that occurred in the past. Consequently, it does not provide an answer to the key question from the point of view of the resilient system: what factors enable one to avoid undesirable events. Active monitoring in OSH management systems means "the ongoing activities which check that hazard and risk preventive and protective measures, as well as the arrangements to implement the OSH management system, conform to defined criteria" (ILO 2001) and can be perceived as an equivalent to monitoring in the concept of the resilience engineering. Therefore, in the both concepts, monitoring should be performed in a proactive way, i.e. before an adverse event occurs. However, the nature of the monitoring itself is different. In OSH management systems, the starting point for the assessment of monitoring results is the organisation's safety health and objectives and the compliance of the activities with the procedures. Meanwhile, resilience engineering is mainly about monitoring the adequacy of OSH management processes to ensure that any emerging dysfunctions are controlled to avoid unexpected adverse events. The difference can be expressed in Zwetsloot's words (2003): "active monitoring in traditional OSH management systems means monitoring whether good things are done, while in the concept of resilience engineering it is whether things are done right".

Anticipating risks in the field of occupational safety and health and taking appropriate measures in advance to prevent accidents at work or industrial accidents does not seem to be something new for enterprises, especially those that have a safety and health management system in place. This issue is primarily addressed in the risk assessment process that is defined as a "combination of the likelihood of an occurrence of a hazardous event and the severity of injury or damage to the health of people caused by this event" (ILO 2001).

Probability means the need to anticipate adverse events in the field of occupational safety and health on the basis of our knowledge and experience. Moreover, in OSH management systems, in order to take a priori appropriate preventive measures, it is necessary to identify risks before purchasing goods and services and signing contracts with subcontractors. Therefore it seems that anticipation in traditional management systems is largely in line with the concept of resilience engineering.

In OSH management systems high priority is given to emergency prevention, preparedness and response arrangements: "These arrangements should identify the potential for accidents and emergency situations, and address the prevention of OSH risks associated with them" (ILO 2001). The performance of these arrangements, i.e. responding, depends on the quality of the performance of the remaining ones and, to be more specific, the knowledge about the realisation of occupational safety and health management processes. However, this is limited by the fact that OSH management systems do not fully introduce learning and monitoring in accordance with the assumptions of the resilience engineering concept. Research on relationship between certified OSH management systems and resilience engineering confirms the introduction of a certified OSH management system alone does not result in the ensuring of a resilient system (Figure 4.5) (Pęciłło 2016).

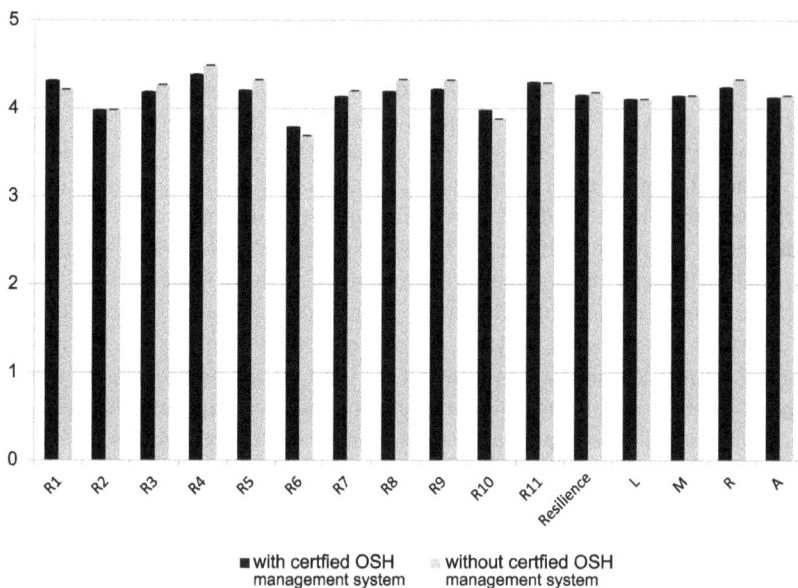

FIGURE 4.5 Resilience engineering level in enterprises with and without certified OSH management systems for particular elements of OSH management system. (reprinted, with permission, from: Pęcillo 2015. *Materiały informacyjne nt. Resilience engineering w zarządzaniu bezpieczeństwem i higieną pracy* [Informative materials on resilience engineering In OSH management], CIOP-PIB, p. 4)Where: R1 is management involvement; R2 is participation of workers; R3 is planning; R4 is responsibility and authority; R5 is training, awareness, and qualifications; R6 is motivation; R7 is communication; R8 is occupational risk management; R9 is preparedness and response to accidents at work and failures; R10 is subcontractors; R11 is monitoring; Resilience (total); L is learning; M is monitoring; R is responding; and A is anticipating.

This gap is to some extent filled by the latest standard on OSH management systems: ISO 45001 (ISO 2018). In the system, as a rule, identification of both risks and opportunities provides the basis for managing safety and health at work. The identification of opportunities, similarly to risks, should be based on past events (learning), current events (monitoring) and future events (anticipation). In a way, thinking towards the concept of resilience engineering is also introduced by moving away from single actions or those repeated only periodically in favour of their continuity and introducing a process approach. According to the ISO 45001 standard an organisation should plan, implement and control the processes needed to meet requirements of the OSH management system, and to take actions ensuring, inter alia, prevention and reduction of undesired effects. This should be done by establishing criteria for the processes and implementing control of the processes in accordance with the criteria, keeping documented information to the extent necessary to have confidence that the processes have been carried out as planned, and, finally, by determining situations where the absence of documented information could lead to deviations from the OSH policy and the OSH objectives.

4.2.3 Trade-offs in Resilience Engineering Concept

4.2.3.1 The Concept of Trade-offs

A trade-off is a situation in which you must choose between or balance two things that are opposite or cannot be had at the same time or something that you do not want but must accept in order to have something that you want (Merriam-Webster Dictionary). Thus, making a trade-off involves a choice and some sacrifices. A good example of a trade-off is using different file formats for music files, as it provides a choice between sound quality and disk space. While the number of dishes on a restaurant menu may be another good example, it illustrates also the trade-off between more efficient quality control and product variability that may attract additional customers (Frei 2006). Similarly, another classical idea is that one must choose between doing something quickly and doing something well. While the idea itself may be questionable in many cases, it reflects one of the most important problems in production and quality management, and, more general, in balancing between effectiveness and other values, including quality and safety of both production process and product.

Trade-offs have been very well recognised and described by numerous sciences. One of the sciences treating managing trade-offs as one of the key ideas is ecology. Ecosystems are complex and dynamic systems of interacting components and both the case of adaptation to habitat and influence of changes resulting from human activity on ecosystems' functioning are among the most highlighted areas of ecological research. Animals "choose" their strengths. Sprint speed in savannah herbivores is correlated with vulnerability to the main predators in the community, as one cannot have opposite adaptations at the same time (Bro-Jørgensen 2013). Human activity makes irreversible changes in natural ecosystems, but, as some activities are highly desirable, some decisions are made to sacrifice the ecological component in favour of the social and economic component (Birge et al. 2014). Damming is necessary for flood protection and positively influences other fields, including support for agriculture and energy production, while, on the other hand, it may greatly reduce hydrological variability and endanger the species adapted to bare sandbar habitats created by pulsing floods (Allen et al. 2018). High agricultural productivity is usually achieved by lowering biodiversity or species richness (Cavender-Bares et al. 2015).

One of the most obvious trade-offs in economy is whether to save or invest money. Saving money is safer and gives relatively small return, but, at the same time, it is as safe as the existence of the given bank. On the other hand, depending on the chosen way of investment, there may be a possibility of high returns, while loss of all the invested money may happen as well. Usually, we may expect a bigger expected return the higher the level of risk.

In management practice, trade-offs are not always as obvious and clear as the examples mentioned above, and very often potentially conflicting goals may stay balanced for long period of time. In such case, tensions may occur when external or internal pressure makes an organisation divert more resources to one goal, while there are not sufficient resources remaining for the other one and, as the result, the goals become imbalanced. As far as quality management is concerned, a good illustration of this problem may be an organisation with a relatively stable production level and a good product quality being achieved with well-performing quality control. Rapid production volume increase may result in worsening quality control process, if it is not prepared and strengthened enough for bigger challenge. Of course, the choice is quite obvious when taking into account intentionally lowering product quality to achieve higher sales by decreasing the price. The cases of trade-offs involving quality include the production–quality, effectiveness–quality, cost–time–quality and price–quality trade-offs, along with many others, and are very well recognised and described, as quality management has been well developed and recognised (Pinker and Shumsky 2000; Collins Dictionary of Business; Dodds et al. 1991; Carmon and Simonson 1998). Many good examples come from strategic choices, as in the case of trade-offs between the exploration of new possibilities and the utilisation of existing opportunities (March 1991) or between diversification and concentration (Boyle et al. 2012).

4.2.3.2   Trade-offs in Resilience Engineering

Trade-offs are one of the major ideas in resilience engineering theory (Woods 2006a). This theory states that the nature of multi-role systems results in the necessity of balancing trade-offs (Woods and Hollnagel 2006). The importance of trade-offs results from the fact that resilience engineering focuses on dynamic processes, which, depending on trade-offs made, may improve or undermine resilience of a system. Such approach makes managing trade-offs one of the key factors for successfully building a safe organisation. It also highlights the importance of all the factors influencing the managing of decision-making processes, which are critical for an organisation to be resilient. Properly defined values, priorities and goals are needed to make the best possible trade-offs (Hoffman and Woods 2011; Hollnagel 2009).

When analysing cases of disturbances of an organisation's activity, it may be easily noticed how the processes are both shaped and affected by specific decisions made at each stage of the processes and each level of an organisation, from strategic decisions made by the highest authorities to the simplest decisions in specific cases made by front workers. This perspective enables one to appreciate the significance of proper managing and making trade-offs for the ability of a system or an organisation to be resilient.

When reading works on resilience engineering it can be noticed that authors usually describe the four cornerstones and resilient systems very theoretically, with use of complicated schemes and charts, while describing the trade-offs usually involves studies of one or a few cases when a decision was made by specific people with little concern for a wider view, as in case of the engineers voting on whether or not to allow the launch of the Challenger space shuttle (Vaughan 1996). This example shows also how vulnerable, and even defenceless, a system may be at the time when a decision is being made and how a wrong choice may affect an organisation, even one as huge as NASA. It may be partly excused by a hardly lack of valid and verified tools developed for the description and management of trade-offs. Consequently, some tools providing frameworks for assessing resilience based on the four cornerstones have been developed. However, they usually omit trade-offs and ignore their role thereof for making and keeping an organisation resilient.

Probably the biggest number of cases of trade-offs come from aviation (and, more widely speaking, from transportation), including the famous US Airways Flight 1549 and its ditching in the Hudson River, as that involves many elements examined by resilience engineering, such as emergency, shortage of resources (including limited time and insufficient information for decision making), risk management, replanning, recovery, and trade-offs. Pariès describes the decision to ditch in the Hudson as the trade-off between "either the Hudson, certainly bad but possibly not catastrophic, or surrounding airports, possibly a happy ending, with minimum damage to the airplane, but almost certainly catastrophic in case of failure of the attempt" (Pariès 2011). The case of postponed repairs and shipyard visits of merchant vessels is good example of sacrificing safety for efficiency (Heese et al. 2013). The same kind of trade-off is presented by "classical" examples of the trade-off from aviation: maintenance technicians examining aircraft. Tjørhom and Aase (2011) describe the case of technicians making trade-offs between keeping the aircraft technically airworthy and keeping to the planned flight schedule and increasing demand on effectiveness and productivity. Another case is the choice between strictly following rules and flexibility coming from common sense, experience and the need to adjust to actual circumstances. According to McDonald, "most commonly the technicians reported that there were better, quicker, even safer ways of doing the tasks than following the manual to the letter" (McDonald 2006), which is illegal and violates the rules. In a broader context, this problem arises from the difference between theory (work-as-imagined) and practice (work-as-done) (Braithwaite et al. 2016) and should be matter of significant concern whenever strict regulations not allowing any flexibility are introduced in given organisation.

Focusing the attention on trade-offs changes the approach to what is important when analysing factors of success, as the factors influencing the decision-making process become the most important. This is somehow an answer to complaints that resilience engineering "has no people in it" (Hunte et al. 2013; Wears 2011) and recalls the need for switching some focus from system to people. It also highlights the role of establishing and prioritising goals and rules and guidelines for decision making and, on the other hand, the necessity of sustainable safety and pro-resilient corporate culture in an organisation. It does not mean that one universal model of resilient culture may be defined and introduced all around the world, at least because of cultural differences (Bracco et al. 2013).

One of the major ideas under resilience engineering is that being resilient means both being prepared and being prepared to be unprepared. According to Pariès, this contradictory idea contains a trade-off between efficiency and flexibility (Pariès 2011).

4.2.3.3 Trade-off Typology in Resilience Engineering

The number and variability of cases involving trade-offs has caused researchers to try to create a typology facilitating discussion and practical use of the theory. Hoffman and Woods (2011) created a list of five major trade-offs, including acute–chronic, efficiency–thoroughness, specialist–generalist, distributed–concentrated and optimality–fragility trade-offs.

An acute–chronic trade-off is made when one has to choose between acute goals, such as efficiency, productivity and effectiveness, and chronic goals, that sometimes are identified in long-term goals and values. The tendency of choosing acute or chronic goals shows whether an organisation sacrifices long-term goals and values to achieve acute goals under pressure to take actions that may be contrary to the safety policy of the organisation. This pressure is very often effective, because, while chronic goals are somehow values which influence on the performance of an organisation cannot be easily seen in the short term and from indicators usually used by management or shareholders for evaluation of performance, acute goals are formulated on the basis of current, constantly changing, policies following on-time analysis of possibilities and needs at the moment and can usually be measured by the indicators used to evaluate performance of an organisation and of specific workers, particularly managerial staff (Woods 2009; Woods 2011).

Contradiction between reduction of thoroughness to improve productivity and demands of safety is reflected by the efficiency–thoroughness trade-off (the ETTO principle). The idea of the trade-offs is that maximising thoroughness and effectiveness at the same time is never possible. There is a choice between spending resources (time and effort) on preparing activity or spending resources on performing activity (Hollnagel 2009). This trade-off is also defined as gap in plans (Hoffman and Woods 2011). Such a definition assumes that effective activity requires using well-worn paths, which results in limiting the possibility of changing the paths or introducing unexpected changes or variations, while thoroughness involves expanding the scope of plans including some ambiguities, variations, amendments and possible decisions on some changes.

Responsibilities and duties in contemporary complex organisations are divided. Various units or elements of a system have their own goals and their own scope of activity. While the goals for specific units are defined on the basis of common goals for the whole organisation, in practice they may be conflicting. This problem is reflected by the specialist–generalist trade-off. According to Woods and Branlat (2011), multiple conflicting goals and the distribution of responsibility among different parts of a system result in the arising of gaps in roles. Consequently, different parts of a system may be co-operative over shared goals and competitive where goals conflict.

The distributed–concentrated trade-off reflects the choice between concentration and distribution of an authority and control (Woods and Branlat 2011). Woods and Wreathall (2008) state that this trade-off results from gaps in progress. To increase the effectiveness and range of action, and to facilitate flexibility and, in some cases, to optimise time and appropriateness of response, responsibility and, consequently, activity are distributed, usually to lower levels of management. On the other hand, such distribution may lead, and very often leads, to fragmentation of the process managed by numerous decision centres, which results in the need to make efforts and to invest additional resources to keep distributed activities coherent and synchronised.

The optimality–fragility trade-off has been well recognised by research of biological and physical systems. The trade-off reflects the problem that optimisation of performance in a given environment results in vulnerability to change or disturbances (Carlson and Doyle 2002; Zhou et al. 2005; Csete and Doyle 2002). For the purpose of resilience engineering theory, Woods and Wreathall (2008) reformulate it to the statement that improving system's optimality may lower the resilience thereof. In other words, it is not possible to optimise and at the same time to improve adaptive capacity (Hoffman and Woods 2011). An optimality–fragility trade-off may be a good starting point to analyse to what extent the strategy of building a resilient organisation may be compatible with lean management. Some authors point out that the introduction of both concepts has a similar motivation and are not mutually exclusive (Rosso and Saurin 2018). Both lean management and resilience engineering seem to be quite compatible, as their common aim is improving the management processes. However, attention should be drawn to the different ways in which this objective can be achieved. In order to achieve lean management objectives, system optimisation, predictability and repeatability are required. Meanwhile, the resilience engineering concept is based on the conviction that a resilient system must be flexible in order to cope with unpredictable situations. This is because only a flexible system, prepared for the appearance of dysfunctions (also as unpredictable as possible), is able to cope with a dynamic environment. To this end, all reserves of resources, including knowledge (both explicit and implicit), are helpful. Meanwhile, lean management strives to minimise resources, which often translates into reduced employment and outsourcing. Such an approach certainly makes it difficult to take action in critical situations (Table 4.2).

Lean management deals, in a great simplification, with the optimisation of the functioning of an organisation in a given state and operating in a given environment; while the task of resilience engineering is to prepare the organisation to avoid or face difficulties, including unpredictable ones. From this point of view, the full benefits of the introduction and execution of lean management principles can be obtained when the organisation operates in relatively stable and predictable conditions. Positive effects of a resilient organisation are revealed in crisis situations, which may lead to accidents at work and catastrophes in unprepared (non-resilient) enterprises.

TABLE 4.2
Comparison of the Basic Concepts of Lean Management and Resilience Engineering

	Lean management	Resilience engineering
Aim	Improvement of effectiveness understood as the ratio of outlays to results	Building capacity to respond to crisis situations
Results	Optimal system	Flexible system
Process to be improved	A process in which there are long downtimes, which do not create added value	A process incapable of responding to unexpected dysfunctions that arise
Resources	Optimised	Striving to maintain reserves

Reprinted, with permission, from: Pęciłło 2019 Zarządzanie według koncepcji resilience engineering i lean management – porównanie w kontekście bezpieczeństwa i higieny pracy. (Management according to resilience engineering and lean management concepts – comparison from OSH perspective) Bezpieczeństwo Pracy. Nauka i Praktyka 2019,3:20–22.

An example of a different approach to trade-off typology was used by authors of Resilience Matrix (Bracco et al. 2013). To facilitate managing trade-offs and enabling providing proper response for the signals that could warn of future threats and opportunities, they describe trade-offs as three dilemmas. Two of those concern signals directly: how to recognise "right" signals demanding response and which actor should intervene in a certain situation. The third one is between resilient culture that potentially should be enhanced to improve potential for making resilient responses and those existing in an organisation. The authors give an example of traditional Italian corporate culture as contradictory to culture that seems to support resilience.

4.2.3.4   Tools for Managing Trade-offs

The key role of making trade-offs implies the necessity to introduce measures to manage the decision-making process. However, the majority of effort is focused on the four cornerstones. Moreover, variability of both trade-offs themselves and the situations in which trade-offs are made in makes development of universal tools for trade-off management more complex. Nonetheless, some tools for trade-off management were prepared. One of them is the Q4-Balance framework that enables one to balance proactive and reactive indicators on both safety and economy by mapping and visualising the tendencies in trade-offs (Woods et al. 2013).

Rigaud and Martin proposed the Resilience Assessment Framework (RAF). This tool integrates trade-offs into a process of resilience assessment and control (Rigaud and Martin 2013). Trade-offs are linked with each of four perspectives from which the system can be looked at: the system perspective links the optimality–fragility trade-off with threats and safety management for the system as a whole; the network and the unit perspectives involve the specialist–generalist and the distributed–concentrated trade-offs; the acute–chronic trade-off is linked with the agents perspective; while the efficiency–thoroughness trade-off is linked with both the unit and the agents perspectives. Based on this analysis the potential impact of trade-offs on each of the four cornerstones may be evaluated (Table 4.3). According to the findings, four kinds of trade-offs affect areas and dimensions that are difficult to measure with quantitative indicators: agents' perceptions, communication between units, and safety culture and barriers. While the Resilience Assessment Framework was prepared for a specific case, it may be a good pattern for developing similar frameworks for other systems and organisations.

4.3   SUMMARY

The experience of the last several decades shows that improving existing technological and organisational solutions, despite constantly growing expenditures, brings relatively small effects, and the applied solutions are still not able to ensure full safety at work. For example, OSH management systems which, despite being carefully designed, have not led to an expected level of improvement in safety and health at work. According to Frick et al. (2000), formal systems and structures play an important role when there is a low level of functioning, but they cease to do so when that level is at least average.

TABLE 4.3
The Potential Impact of Trade-offs on the Four Cornerstones

Trade-off	Cornerstone			
	Capacity to respond	Capacity to monitor	Capacity to anticipate	Capacity to learn
Acute–chronic	Agent's perceptions (1) functioning of the system, (2) evaluation of the situations, (3) response plan, (4) adaptation	(1) indicators, (2) measurement, (3) criticality of variability in indicators	Ability to identify: (1) consequences of change and innovation, (2) new threats, (3) opportunities	(1) relevancy of the situations, (2) identifying a diversity of lessons, (3) ability to learn lessons
Efficiency–thoroughness	Availability of time, knowledge, information and resources (1) detecting an abnormal situation, (2) recognising the situation, (3) considering the criticality of the situation and decision on responding, (4) responcing	(1) collecting data, (2) evaluating indicators, (3) analysing indicators	(1) identifying a change and innovation, (2) managing a change, (3) analysing risk and opportunities	(1) studying situations, (2) learning from the results of investigations
Specialist–generalist Distributed–concentrated	Communication capacity between units			
Optimality–fragility	Safety culture. Safety barriers.			

Based on (Rigaud and Martin 2013), pp.118–120.

Hence the focus on finding new, alternative solutions, both in terms of the approach to safety itself and in terms of various methods and tools supporting the management of safety and health at work. An example of a new approach to safety can be the idea of Safety-2 created by Hollnagel (2014), and an attempt to produce an alternative approach to safety management and finding a new place for safety in the strategic view of the organisation as a whole is resilience engineering, which, to a certain extent, places safety issues as key to the proper functioning of an organisation as a whole at the strategic level.

The relatively high level of safety and health at work in today's modern organisations often causes them to forget about the importance of the problem on a daily basis. Often these organisations only remember this when something serious actually happens. Then managers find out that accidents at work involve measurable costs and lost profits. Meanwhile, even a cursory analysis of catastrophes indicates that in most cases the inevitability of a catastrophe seemed obvious or at least could be predicted. For example, in Fukushima, the wave height that threatened the power plant was known; the Rana Plaza building was built and operated contrary to the construction plan; in the case of the Challenger shuttle crash, the decision to take off was made against known safety facts. Resilience engineering is the answer to such threats. It also recalls the need to constantly adapt the management of safety management tools and solutions and to be prepared for continuous safety challenges.

As stated above, the huge improvement in working conditions over the last century and the consequent reduction in the number of accidents can mean that occupational safety issues are not seen as requiring real commitment in an organisation's day-to-day work, but only as a set of obligations to be fulfilled in order to meet legal requirements. The same is true of OSH management systems, which in practice are often only treated as another set of procedures to be followed, which are of minor importance. On the one hand, such an approach may result in inadequate taking of OSH activities and, on the other hand, it does not support at all the developing of a proper culture in an organisation, which, in the light of many studies, is essential for creating and maintaining a high level of OSH.

An additional problem – from the point of view of organisation management – is the issue of senior management perceiving the importance of safety when making trade-offs. When managers face a choice, they almost always choose acute goals. In addition to the hierarchy of values and the importance of individual management objectives and areas, one of the reasons for that is the simple fact that, for example, for the assessment of the current and forecast financial situation or sales and production volumes, there is a set of indicators in each organisation, the values and changes of which over time can be relatively easily defined. These indicators are also used to formulate strategic and operational objectives and to account for the proper introduction of those objectives.

At the same time, organisations rarely, if at all, use any tools to link the level of OSH management to strategic management assessment. In addition to defining such links, it is necessary to provide managers with the tools to properly assess the functioning of an OSH management system and diagnose its malfunctions before the unwanted effects of these errors, in the form of accidents, occur. It is also important that the system for measuring the functioning of an organisation should enable one not only to make an assessment of the existing situation, but also, on the basis of the information obtained, to anticipate future events and, as a result, to make the right strategic decisions, because "managers need as much information and knowledge about future events as possible in order to start a chain of anticipation feedback, thanks to which it is possible to avoid an error". (Kuc 2006). An example of a method of assessing the effectiveness and efficiency of an organisation meeting these conditions is the Balanced Scorecard, which enables one to convert an enterprise's strategy into a system of measures that can be used to comprehensively assess the efficiency of the OSH management system, and thus assess the implementation of the enterprise's global strategy in this area.

REFERENCES

Allen, C. R., H. E. Birge, D. G. Angeler, et al. 2018. Quantifying uncertainty and trade-offs in resilience assessments. *Ecology and Society* 23(1):3. https://www.ecologyandsociety.org/vol23/iss1/art3/(accessed July 23, 2019).

AS/NZS (Australian/New Zealand Standard™). 1997/2001. AS/NZS 4804:1997 and AS/NSZ 4804:2001. Occupational health and safety management systems – General guidelines on principles, systems and supporting techniques.

Azadeh, A., R. Yazdanparast, S. A. Zadeh, and A. E. Zadeh. 2017. Performance optimization of integrated resilience engineering and lean production principles. *Expert System with Application* 84:155–170.

Azour, F., H. E. Moussami, S. Dahbi, and L. Ezzine. 2017. Integration of health and safety at work and environment perspectives in the balanced scorecard. In: *Proceedings of the International Conference on Industrial Engineering and Operations Management Rabat, Morocco*, April 11–13., 2017. https://pdfs.semanticscholar.org/aed8/888f86f0cde975898ef717483a938135065d.pdf (accessed September 18, 2019).

Birge, H. E., C. R. Allen, R. K. Craig, et al. 2014. Social-ecological resilience and law in the Platte River Basin. *Idaho Law Review* 51(1):229–256.

Bourguignon, A., V. Malleret, and H. Nørreklit. 2004. The American balanced scorecard versus the French tableau de bord: The ideological dimension. *Management Accounting Research* 15(2):107–134.

Boyle, P., L. Garlappi, R. Uppal, and T. Wang. 2012. Keynes meets Markowitz: The trade-off Between familiarity and diversification. *Management Science* 58(2):253–272.

Bracco, F., T. F. Piccino, and G. Dorigatti. 2013. Turning variability into emergent system: The resilience matrix for providing strong responses to weak Sygnals. In: Herrera, I., Schraagen, J. M., van der Worm, J., Woods, D. (Eds.). *Proceedings: 5th REA Symposium: Managing Trade-offs*. Resilience Engineering Association, 23–28. http://www.resilience-engineering-association.org/wp-content/uploads/2016/09/Frontpage-REA5SYM-proceedings-030916.pdf (accessed September 18, 2019).

Braithwaite, J., R. Wears, and E. Hollnagel. 2016. *Resilient Health Care III: Reconciling Work-As-Imagined with Work-As-Done*. Farnham, UK: Ashgate.

Bro-Jørgensen, J. 2013. Evolution of sprint speed in African savannah herbivores in relation to predation. *Evolution: International Journal of Organic Evolution* 67(11):3371–3376.

Brown, M. G. 2006. *Keeping Score. Using the Right Metrics to Drive World-Class Performance*. New York: Productivity Press.

BSI (British Standard Institution). 2007. BS OHSAS 18001: Occupational health and safety management: Requirement 2007.

BSI (British Standard Institution). 2004. BS 8800: 2004 Guide to occupational safety and health management systems.

Carlson, J. M., and J. Doyle. 2002. Complexity and robustness. *Proceedings of the National Academy of Sciences of the United States of America* 99(Supplement 1):2538–2545.

Carmon, Z., and I. Simonson. 1998. Price–quality trade-offs in choice versus matching: New insights into the prominence effect. *Journal of Consumer Psychology* 7(4):323–343.

Cavender-Bares, J., S. Polasky, E. King, and P. Balvanera. 2015. A sustainability framework for assessing trade-offs in ecosystem services. *Ecology and Society* 20(1):17. https://ww w.ecologyandsociety.org/vol20/iss1/art17/ (accessed September 18, 2019).

Collins Dictionary of Business, 3rd ed. S.v. "price-quality trade-offs." https://financial-diction ary.thefreedictionary.com/price-quality+trade-offs (accessed September 18, 2019).

Csete, M. E., and J. C. Doyle. 2002. Reverse engineering of biological complexity. *Science* 295(5560):1664–1669.

Debnath, A., J. Roy, K. Chatterjee, and S. Kar. 2018.Measuring corporate social responsibility based on fuzzy analytic networking process-based balance scorecard model. *International Journal of Information Technology and Decision Making* 17(04):1203–1235.

DJSI Index Family. www.sustainability-index.com (accessed January 17, 2020).

Dodds, W. B., K. B. Monroe, and D. Grewal. 1991. Effects of price, brand, and store information on buyers' product evaluations. *Journal of Marketing Research* 28(3):307–319.

Edvinsson, L., and M. S. Malone. 1997. *Intellectual Capital: Realizing Your Company's True Value by Finding Its Hidden Brainpower*. New York: Harper Business.

Frei, F. X. 2006. Breaking the trade-off between efficiency and service. *Harvard Business Review* 84(11):93–101. https://hbr.org/2006/11/breaking-the-trade-off-between-effic iency-and-service (accessed September 18, 2019).

Frick, K., P. L. Jensen, M. Puinlan, and I.Wiltnagen. 2000. *Systematic Occupational Health and Safety Management. Perspectives on an International Development*. Amsterdam: Pergamon.

Friedag, H. R., and W. Schmidt. 2004. *My Balanced Scorecard*. Munich: Haufe Verlag.

Griffin, R. W. 2011. *Fundamentals of Management*. Boston, MA: Cengage Learning.

Heese, M., W. Kallus, and Ch. Kolodej. 2013. Assessing behavior towards organizational resilience in aviation. In: *Proceedings: 5th REA Symposium: Managing Trade-offs*. Resilience Engineering Association, 67–74. http://www.resilience-engineering-associ ation.org/wp-content/uploads/2016/09/Frontpage-REA5SYM-proceedings-030916.p df (accessed September 18, 2019).

Heinrich, H. W. 1941. *Industrial Accident Prevention*. New York-London: McGraw-Hill Book Company.

Hoffman, R. R., and D. D. Woods. 2011. Beyond Simon's slice: Five fundamental trade-offs that bound the performance of macrocognitive work systems. *Intelligent Systems, IEEE* 26(6):67–71.

Hollnagel, E. 2006. Resilience – The challenge of the unstable. In: Hollnagel, E., Woods, D., Leveson, N. (Eds.). *Resilience Engineering: Concepts and Precepts*. UK: Ashgate Publishing Ltd. Aldershot.

Hollnagel, E. 2009. *The ETTO Principle: Efficiency-Thoroughness Trade- off, Why Things That Go Right Sometimes Go Wrong*. Farnham, Surrey, UK: Ashgate.

Hollnagel, E. 2011. Prologue: The scope of resilience engineering. In: Hollnagel, E., Pariès, J., Woods, D. D., Wreathal, J. (Eds.). *Resilience Engineering in Practice*. Farnham, UK: Ashgate.

Hollnagel, E. 2014. *Safety-I and Safety-II: The Past and the Future of Safety Management*. Farnham, UK: Ashgate.

Hollnagel, E., C. P. Nemeth, and S. W. P. Dekker. (Eds.). 2008. *Resilience Engineering Perspectives, Volume 1: Remaining Sensitive to the Possibility of Failure*. Farnham, UK: Ashgate.

Hunte, G. S., R. L. Wears, and C. C. Schubert. 2013. Structure, agency and resilience. In: Herrera, I., Schraagen, J. M., van der Worm, J., Woods, D. (Eds.). *Proceedings: 5th REA Symposium: Managing Trade-offs.* Resilience Engineering Association, 49–54. http://www.resilience-engineering-association.org/wp-content/uploads/2016/09/Frontpage-REA5SYM-proceedings-030916.pdf (accessed September 18, 2019).

ILO (International Labour Organisation). 2001. ILO-OSH 2001: Guidelines on occupational safety and health management systems. https://www.ilo.org/wcmsp5/groups/public/---ed_protect/---protrav/---safework/documents/normativeinstrument/wcms_107727.pdf (accessed September 18, 2019).

Ingalls, T. S. 1999. Using scorecards to measure safety performance. *Professional Safety* December:23–27.

ISO (International Organization for Standardization). 2018. ISO 45001: Occupational health and safety management systems: Requirements with guidance for use.

Kaplan, R. S., and D. P. Norton. 1992. The balanced scorecard – Measures that drive performance. *Harvard Business Review* 1(1):71–79. https://steinbeis-bi.de/images/artikel/hbr_1992.pdf (accessed September 18, 2019).

Kaplan, R. S., and D. P. Norton. 1996. *The Balanced Scorecard: Translation Strategy into Action.* Boston, MA: Harvard Business School Press.

Kuc, B. R. 2006. *Kontroling narzędziem wczesnego ostrzegania.* Warszawa: Wydawnictwo Menedżerskie.

Kueng, P. 2000. Process performance measurement system: A tool to support process-based organizations. *Total Quality Management* 1(1):67–85.

Langhoff, T. 2002. *Ergebnisorientierter Arbeitsschutz – Bilanzierung und Perspektiven eines. Schriftenreihe der Bundesanstalt für Arbeitsschutz und Arbeitsmedizin. Forschung* (Fb 955). Dortmund: Bundesanstalt fur Arbeitsschutz und Arbeitsmedizin.

March, J. G. 1991. Exploration and exploitation in organizational learning. *Organization Science* 2(1):71–87. https://www3.nd.edu/~ggoertz/abmir/march1991.pdf (accessed September 18, 2019).

McDonald, N. 2006.Organisational resilience and industrial risk. In: Hollnagel, E., Woods, D., Leveson, N. (Eds.). *Resilience Engineering: Concepts and Precepts.* Farnham, UK: Ashgate, 155–180.

Mearns, K., S. Whitaker, R. Flin, R. Gordon, and P. O'Connor. 2003. Benchmarking human organizational factors in offshore safety. In: *Factoring the human into safety: Translating research into practice. Volume 1 Research Report 059.* Health and Safety Executive. http://www.hse.gov.uk/research/rrpdf/rr059.pdf (accessed September 18, 2019).

Merriam-Webster Dictionary. https://www.merriam-webster.com/dictionary/trade-off (accessed September 18, 2019).

Neely, A., C. Adams, and M. Kennerley. 2002. *The Performance Prism. The Scorecard for Measuring and Managing Business Success.* London: Prentice Hall.

Olve, N.-G., C.-J. Petri, J. Roy, and S. Roy. 2003. *Making Scorecards Actionable: Balancing Strategy and Control.* Chichester: John Wiley & Sons.

Olve, N.-G., J. Roy, and M. Wetter. 2002. *Performance Drivers: A Practical Guide to Using Balanced Scorecard.* Chichester: John Wiley & Sons.

Panayiotou, N., K. Aravosis, and P. Moschou. 2007. Towards a balanced CSR performance. In: *Measurement Framework Proceedings of the International Conference of Environmental Management, Engineering, Planning and Economics,* Greece: Skiathos Island, June 24–28, 2353–2354. http://arvis.simor.ntua.gr/Attachments/Publications/Conferences/meperilipsistapraktika/7.8.4_TOWARDS%20A%20BALANCED%20CSR%20PERFORMANCE%20MEASUREMENT%20FRAMEWORK.pdf (accessed September 18, 2019).

Pariès, J. 2011. Lessons from Hudson. In: Hollnagel, E., Pariès, J., Woods, D. D., Wreathal, J. (Eds.). *Resilience Engineering in Practice. A Guidebook*. Farnham, Surrey: Ashgate, 9–27.

Pęciłło, M. 2008. Assessing effectiveness of occupational safety and health management. In: Weiss, E., Godlewska, M., Bitkowska, A. (Eds.). *Nowe Trendy i Wyzwania w Zarządzaniu. Koncepcje zarządzania*. Warszawa: Vizja Press & IT, 429–439.

Pęciłło, M. 2015. *Materiały informacyjne nt. Resilience engineering w zarządzaniu bezpieczeństwem i higieną pracy*. (Informative materials on resilience engineering In OSH management). Warszawa: CIOP-PIB.

Pęciłło, M. 2016. The resilience engineering concept in enterprises with and without occupational safety and health management systems. *Safety Science* 82:190–198.

Pęciłło, M. 2019. Zarządzanie według koncepcji resilience engineering i lean management – Porównanie w kontekście bezpieczeństwa i higieny pracy. (Management according to resilience engineering and lean management concepts – Comparison from OSH perspective). *Bezpieczeństwo Pracy – Nauka i Praktyka* 3:20–22.

Petersen, D. 2000. Safety Management 2000: Our strengths and weaknesses. *Professional Safety* January:16–19.

Petersen, D. 2001. The safety scorecard: Using multiple measures to judge safety system effectiveness. *Occupational Hazards* 5:54–57.

Pinker, E. J., and R. A. Shumsky. 2000. The efficiency-quality trade-off of cross trained workers. *Manufacturing and Service Operations Management* 2(1):32–48.

PKN (Polski Komitet Normalizacyjny, Polish Committee for Standardization). 2001. PN-N-18004:2001 Occupational safety and health management systems. Guidelines.

PKN (Polski Komitet Normalizacyjny, Polish Committee for Standardization). 2004. PN-N-18001:2004 occupational safety and health management systems. Requirements.

Rigaud, E., and C. Martin. 2013. Considering trade-offs when assessing resilience. In: Herrera, I., Schraagen, J. M., van der Worm, J., Woods, D. (Eds.). *Proceedings: 5th REA Symposium: Managing Trade-offs*. Resilience Engineering Association, 115–120. http://www.resilience-engineering-association.org/wp-content/uploads/2016/09/Frontpage-REA5SYM-proceedings-030916.pdf (accessed September 18, 2019).

Rosso, C. B., and T. A. Saurin. 2018. The joint use of resilience engineering and lean production for work system design: A study in healthcare. *Applied Ergonomics* 71:45–56.

Rummler, G. A., and A. P. Brache. 1995. *Improving Performance: How to Manage the White Space on the Organization Chart*. San Francisco, CA: Jossey-Bass.

Shannon, H. S., J. Mayer, and T. Haines. 1997. Overview of relationship between organizational and workplace factors and injury rates. *Safety Science* 26(3):201–217.

Tjørhom, B., and K. Aase. 2011. The art of balance: Using upward resilience traits to deal with conflicting goals. In: Hollnagel, E., Pariès, J., Woods, D. D., Wreathal, J. (Eds.). *Resilience Engineering in Practice*. Farnham, Surrey: Ashgate, 157–170.

Toellner, J. 2001. Improving safety and health performance: Identifying & measuring leading indicators. *Professional Safety* 9:42–47.

Tuomela, T.-S. 2000. *Customer Focus and Strategic Control: A Constructive Case Study of Developing a Strategic Performance Measurement System at FinABB*. (Turun kauppakorkeakoulun julkaisuja. Sarja D-2.

Vaughan, D. 1996. *The Challenger Launch Decision: Risk Technology, Culture and Deviance at NASA*. Chicago, IL: University of Chicago Press.

Wears, R. L. 2011. *Exploring the Dynamics of Resilient Performance: Business Administration*. Paris: École Nationale Supérieure des Mines de Paris. https://pastel.archives-ouvertes.fr/pastel-00664145/document (accessed September 18, 2019).

Witcher, B., and V. S. Chau. 2007. Balanced Scorecard and Hoshin Kanri: Dynamic capabilities for managing strategic fit. *Management Decision* 45(3):518–538.

Woods, D. D. 2006a. Essential characteristics of resilience. In: Hollnagel, E., Woods, D., Leveson, N. (Eds.). *Resilience Engineering: Concepts and Precepts*. Aldershot: AshgatePub Co, 21–34.

Woods, D. D. 2006b. Resilience engineering: Redefining the culture of safety and risk management. *HFES Bulletin* 49(12):1–3. http://ordvac.com/soro/library/Aviation/Aviation%20Safety/General%20Safety%20Articles/resilience%20engineering%20bulletin.pdf (accessed September 18, 2019).

Woods, D. D. 2009. Escaping failures of foresight. *Safety Science* 47(4):498–501.

Woods, D. D. 2011. Resilience and the ability to anticipate. In: Hollnagel, E., Pariès, J., Woods, D. D., Wreathall, J. (Eds.). *Resilience Engineering in Practice*. A Guidebook. Farnham, UK: Ashgate.E, 121–125.

Woods, D. D., and M. Branlat. 2011. How human adaptive systems balance fundamental trade-offs: Implications for polycentric governance architectures. In: Hollnagel, E., Rigaud, E., Besnard, D. (Eds.). *Proceeding of the Fourth Resilience Engineering Symposium*. Paris: Presses des Mines, 277–283.

Woods, D. D., and E. Hollnagel. 2006. *Joint Cognitive Systems: Patterns in Cognitive Systems Engineering*. Boca Rayton, FL: Taylor&Francis/CRC Press.

Woods, D. D., and J. Wreathall. 2008. Stress-strain plot as a basis for addressing system resilience. In: Hollnagel, E., Nemeth, C. P., Dekker, S. W. A. (Eds.). *Resilience Engineering Perspectives: Remaining Sensitive to the Possibility of Failure*. Adelshot, UK: Ashgate, 143–158.

Woods, D. D., I. Herrera, M. Branlat, and R. Woltjer. 2013. Identifying imbalanced in a portfolio of safety metrics: The Q4-Balanace framework for economy-safety tradeoffs. In: Herrera, I., Schraagen, J. M., van der Worm, J., Woods, D. (Eds.). *Proceedings: 5th REA Symposium: Managing Trade-offs: Resilience Engineering Association*, 149–155. http://www.resilience-engineering-association.org/wp-content/uploads/2016/09/Frontpage-REA5SYM-proceedings-030916.pdf (accessed September 18, 2019).

Zhou, T., J. Carlson, and J. Doyle. 2005. Evolutionary dynamics and highly optimized tolerance. *Journal of Theoretical Biology* 236(4):438–447 http://web.physics.ucsb.edu/~complex/pubs/JTB_236.pdf (accessed September 18, 2019).

Zwetsloot, G. I. J. M. 2003. From management systems to corporate social responsibility, Journal of Business Ethics. Special Issue on *Corporate Social Responsibility* 44(2):201–207.

5 OSH Management from the Corporate Social Responsibility Perspective

Zofia Pawłowska

CONTENTS

5.1 OSH-RELATED ISSUES OF SOCIAL RESPONSIBILITY

Over the past decades numerous concepts and definitions of CSR have been developed by different authors which recognise CSR as obligation, initiative, concept, approach, corporate commitment, process, business or management practice (Mullerat 2011). Although different phrases have been used to define the CSR, all the definitions are, generally speaking, consistent (Dahlsrud 2006). They all emphasise that socially responsible organisations should integrate economic, social, ethical and environmental issues in organisations' policies and actions while taking into consideration stakeholders' needs and expectations.

For years, improving the quality of life has been considered one of the most important objectives of CSR. Hopkins (2005) stated that the aim of social responsibility is to create higher and higher standards of living, while preserving the profitability of the corporation. He also emphasised that CSR is concerned with treating the stakeholders of the firm ethically, in a socially responsible manner. Since in each organisation employees are one of the most important stakeholder groups and are closely integrated with the organisation, the implementation of the CSR principles means that the employees are treated fairly and equitably (Sowden and Sinha 2005; Moir 2001).

As early as in 1999, the World Business Council for Sustainable Development recognised improvement in the quality of life of employees, their families, the local community and society as a whole as an important result of socially responsible behaviour of business and pointed out that CSR is about helping to meet people's needs (Micheal 2003). As quality of work has a significant impact on quality of life, any socially responsible organisation should strive to improve it, inter alia by improving working environment, including working conditions, training, career prospects, health and well-being, etc. (Clark 1998).

Shaping the working environment to ensure the safety, health and well-being of workers is the domain of safety and health at work, which is understood as a multidisciplinary and comprehensive approach that considers individual's physical, mental and social well-being, general health and personal development (WHO 1994). The close link between CSR and occupational safety and health was highlighted years ago in EU-OSHA publications (Zwetsloot and Starren 2004; Pawłowska 2013).

The internationally accepted definition of social responsibility was provided in 2010, in the ISO 26000 standard, which is intended to help organisations with understanding and implementing social responsibility. Social responsibility has been defined as the responsibility of an organisation for the impacts of its decisions and activities on society and the environment through transparent and ethical behaviour (ISO 2010) which:

- contributes to sustainable development, including health and the welfare of society;
- takes into account the expectations of stakeholders;
- is in compliance with applicable law;
- is consistent with international norms of behaviour;
- is integrated throughout the organisation and practiced in its relationships.

In the standard core subjects of an organisation's social responsibility are presented which include: organisational governance, human rights, labour practices, the environment, fair operating practices, consumer issues, and community involvement and development. For each of these subjects basic issues of social responsibility were defined and actions recommended in relation to the issue proposed. The list of these issues may (and surely will) evolve over time to reflect the evolution of social and environmental expectations. Issues of corporate social responsibility related to OSH are primarily addressed in the two core subjects of social responsibility: work practices and human rights.

As defined by the ISO 26000 standard, the labour practices of an organisation are practices relating to work performed within, by or on behalf of the organisation, including sub-contracted work. Social responsibility issues in this area may concern: the recruitment and promotion of workers, disciplinary and grievance procedures; the transfer and relocation of workers, termination of employment; training and development of skills; health, safety and industrial hygiene; and any policy or practice affecting conditions of work, in particular working time and remuneration. The OSH-related issues of social responsibility fall primary in the group titled "health, safety and industrial hygiene" but can be also found among issues presented in relation to human development and workplace training.

From the point of view of social responsibility, safety and health protection at work means not only prevention of harm to health caused by working conditions but also the promotion and maintenance of the highest degree of physical, mental and social well-being of workers. The actions recommended to deal with the issues related to safety and health protection include, inter alia (ISO 2010):

- introducing principles of OSH management and developing, implementing and maintaining an occupational health and safety policy;
- ensuring the participation of workers in OSH-related activities;
- recognising and respecting the rights of workers by ensuring, inter alia, timely, full and accurate information concerning health and safety risks, the best practices used to address these risks and possibility to be consulted on all aspects of their safety and health at work;
- identification and management of occupational risks when taking into account the diversity of the workforce and necessity to adapt working environment to the physical and psychological possibilities and needs of workers;
- striving to eliminate psychosocial hazards in the workplace;
- recording and investigating all safety incidents and health problems in order to minimise or eliminate them;
- providing the safety equipment needed, including personal protective equipment;
- providing adequate training to all personnel on all relevant matters;
- communicating the requirements of the procedures adopted to ensure that workers comply with them; and
- providing equal safety and health protection for part-time and temporary workers, as well as sub-contracted workers.

A large part of these actions is a duty of companies resulting from the provisions of law. Others may be introduced on a voluntary basis in the framework of safety and health management system.

According to ISO 26000, human development includes the process of enhancing people's choices through the development of their abilities and independence. Promoting human development in organisations can be achieved by combating discrimination, maintaining a work–life balance, promoting health and well-being, increasing the diversity of workers employed, and increasing workers' capacity and employability. The actions recommended to deal with these issues are usually thought to be a part of human resources management; however, they also fall within the scope of OSH management. In an attempt to support human development in the framework of OSH management, an organisation may, inter alia:

- ensure training for all workers in occupational risk assessment and safety and health management, as well as in social responsibility issues, to the extent necessary;
- evaluate workers' performance taking into account their attitudes and involvement in OSH and socially responsible activities;
- introduce a system motivating workers' involvement in safety and health and socially responsible actions;
- promote workers' health, e.g. by promoting healthy lifestyles or ensuring access to free recreation for workers and their families;
- provide rehabilitation for workers who have suffered an accident or illness as a result of work and training to enable them to acquire new skills.

OSH is also relevant to another core social responsibility subject: the human rights. In order to respect these rights, every organisation should counteract discrimination, which may be manifested by unreasonable differentiation, exclusion or preference and, consequently, unequal treatment. Ensuring equal opportunities and combating discrimination is important in terms of fundamental rights at work. In striving to respect these rights, employment policies and practices, pay, working conditions, access to training and promotion opportunities, and termination of employment contracts should take into account only job-related requirements and be free from discrimination (e.g. on the basis of gender, religion, national origin, political opinion, age or disability). It is also necessary to prevent harassment in the workplace and to provide protection and support for vulnerable groups. To deal with these issues, in the framework of OSH the following may be considered, inter alia:

- taking into account the health status and capabilities of workers when assessing occupational risks and taking preventive measures;
- introducing programmes/measures targeted at specific groups of workers (e.g. older or young workers, people with disabilities, migrant workers, etc.);
- action addressing psychosocial risks to reduce stress and eliminate mobbing and workplace bullying.

To introduce CSR, each organisation needs processes, systems, structures, or other mechanisms that make it possible to apply the principles and practices of CSR. They provide organisational governance to support an organisation with making and implementing decisions that will enable it to achieve its goals. Organisational governance can include both formal management mechanisms, based on defined structures and processes, and informal mechanisms, based on the organisation's culture and values, which are influenced by the people in charge of the organisation.

In the context of social responsibility, organisational governance plays the role of both a core subject and a method of increasing an organisation's ability to implement socially responsible practices in each of the other core subjects. This double characteristic of organisational governance results from the need to apply an effective decision-making system based on the principles of responsibility, transparency, ethical behaviour, respect for stakeholders' expectations and legal compliance. An important role in the development of organisational governance may be played by management systems, including the OSH management system that can support the development of organisational structures and the leadership and participation of workers in occupational health and safety management.

5.2   DEVELOPING OSH MANAGEMENT SYSTEMS TO IMPLEMENT SOCIAL RESPONSIBILITY

To support implementation of the CSR, an OSH management system should be able to deal with issues of social responsibility related to OSH. The voluntary OSH management systems are usually implemented with the aim of improving safety, the health and well-being of workers and are expected to be an effective tool to implement legal requirements in the area of OSH. They should also support implementation of the CSR good practices (Zwetsloot and Starren 2004; Montero et al. 2009). However, there is no clear evidence from research that OSH management systems can achieve their goals (Robson et al. 2007). A lot of factors negatively influencing the effectiveness of these systems have been identified by research. Unsatisfactory commitment from top management and a lack of participation of workers in OSH management are commonly considered to be the most important (Gallagher et al. 2003; Hale et al. 2010; Podgórski 2005; Frick 2011). Another issue is the focus on documentation, which leads to increases in bureaucratisation and may result in difficulty in identifying real OSH problems, difficulty in predicting unexpected events, introducing bureaucratic accountability, hampering of innovation, etc. (Dekker 2014).

Hasle and Zwetsloot (2011) pointed out that a management system is only a tool, and the effects of its use may vary depending on the purposes for which it is used. This tool, used in a wrong way, can mask but not solve OSH problems (Rocha 2010). In this context, the question arises of whether an OSH management system can really be an effective tool for socially responsible OSH management. Zwetsloot (2003) states that the traditional management systems, with their focus on rational control (doing things right), can only be of limited use in the development of CSR. This has been confirmed by a study performed in 60 large and medium-sized companies with the aim of determining whether the implementation of a certified OSH management system positively influences the socially responsible initiatives and actions in the area of OSH (Pawłowska 2011). The actions evaluated included, inter alia, acknowledgement and promotion of corporate social responsibility in a company's policies and strategies, planning CSR-related actions, psychosocial risks assessment and management, health promotion and rehabilitation, supporting personal development, and motivating workers to participate in actions related to OSH and environment protection. The results show that there are no significant differences in the degree of implementation of the socially responsible actions in companies with and without certified management systems. Moreover, there are no differences in the implementation of measures relating to the psychosocial working environment in both groups of companies. This confirms that the impact of certified occupational health and safety management systems on softer psychosocial issues in the working environment may be questioned and the complexity of psychosocial working environment issues seems to disappear in the case of OSH standards (Hohnen et al. 2014). This also shows that adapting management systems to a changing world of work in which new risks are constantly emerging is becoming increasingly challenging.

To deal with challenges in the changing world of work organisations need to focus on a collaborative approach with the aim of identifying and managing workplace risks and ensuring the integration of OSH into everyday management (Gallagher and Underhill 2012). Incorporating CSR issues in OSH management may be supportive with regard to introducing such an approach and contribute to developing of OSH management systems (Hasle and Zwetsloot 2011). A CSR-inspired approach to OSH management may assist organisations with dealing with psychosocial issues as well as other issues of social responsibility and risks arising in the changing world of work. CSR can also contribute to integration of safety and health into business processes and to developing a strategic approach to OSH which combines the rational logic of prevention and safety management systems with ethical or value-driven approaches (Jain et al. 2018).

An opportunity to improve existing OSH management systems in order to increase their effectiveness with regard to introducing social responsibility seems to be a new standard with requirements and guidelines for OSH management systems, ISO 45001. For organisations implementing and maintaining occupational health and safety management systems, the new standard can be a source of inspiration for how to deal with new challenges, taking into account the needs and expectations of stakeholders. The new standard points to the need to identify the issues covered by OSH management based on an analysis of the organisation's environment (context) and work activities performed for the organisation. The organisation's context includes factors in its external environment, such as demographic, economic, political and legal, socio-cultural, socio-cultural, technical, and natural factors, as well as factors in its internal environment. The internal environment of an organisation consists primarily of the organisational governance, including decision-making processes (formal and informal), organisational structure, responsibilities and powers, policy, objectives and plans aimed at achieving them, resources (including available knowledge and competences), processes, systems, and technologies, as well as working conditions and values recognised by workers. When analysing the context of an organisation, the interests of its stakeholders, i.e. its employees and various external organisations dealing with labour protection issues, should also be taken into account. In this way it is possible to identify and better understand the issues, positive and negative, that the organisation needs to consider in establishing its OSH management system, including the relevant issues of social responsibility in the area of OSH. This should make it possible to adapt a system to the needs of the changing world of work and ensure a socially responsible response to the challenges it poses. One of these challenges is change in the way we live and work linked to developing new technologies and trends emerged, such as the Internet of Things (IoT), robotics, virtual reality (VR) and artificial intelligence (AI). These changes, also called "the fourth industrial revolution" or "Industry 4.0", are the reason that previous approaches to preventive management of workplace health and safety are at risk (Badri et al. 2018). At the same time as creating new threats, new technologies also provide opportunities for better protection of workers, including through possibility to apply ICT solutions in occupational safety and health (OSH) topics. Smart PPE, ambient intelligence and the Internet of Things technologies may make possible the introduction of a new paradigm of OSH risk management consisting of real-time risk assessment and the ability to monitor the risk level of each worker individually (Podgórski et al. 2016).

As important as technological change is demographic change, in particular the ageing of the population. Recognising the ageing of the workforce as an important issue and aiming to maintain work ability, an organisation contributes to the achieving of sustainable development goals, which include the promotion of sustainable, balanced and inclusive growth, full and productive employment and decent work for all (United Nations 2015). To deal with the issue it is necessary to consider factors affecting the work ability in the scope of an OSH management system.

Of course, it cannot be argued that the approach to health and safety management presented in the new ISO standard will make this management socially responsible in every organisation implementing an OSH management system. People, their competences and motivations decide how the postulates contained in it will be introduced. If the only purpose of adjusting to the requirements of the new standard is to certify the system, there will be no changes in the approach to OSH management. An additional opportunity to better understand the relationship between OSH management systems and social responsibility is the newly developed GRI-403 guideline (GRI 2018) which provides guidance with regard to reporting on social responsibility in the area of OSH in a manner consistent with the new ISO standard for OSH management systems.

5.3 INCORPORATING AGE MANAGEMENT INTO SOCIALLY RESPONSIBLE OSH MANAGEMENT

5.3.1 Understanding Age Management and Its Objectives

The term "age management" refers usually to the various dimensions by which human resources are managed within organisations, with an explicit focus on ageing and, also, more generally, to the overall management of the ageing workforce (Walker 1997; TAEN 2007). Age management is based on implementing measures that combat age barriers and/or promote age diversity, with the aim of creating an environment in which individual workers are able to achieve their potential without being disadvantaged by their age (Walker 1998). Years ago, Naegele and Walker (2006) outlined the following dimensions of age management: job recruitment, learning, training and lifelong learning, career development, flexible working time practices, health protection and promotion, workplace design, redeployment, employment exit and transition to retirement, and comprehensive approaches. They believed that age management means a set of age-aware HR practices that can be applied to any of the above dimensions, in entire organisations. On the basis of a literature review concerning ageing workers and human resources management, Aaltio et al. (2011) concluded that age management is usually presented in the literature as the collection of 'best' HR practices which are implemented with the aim of maintaining and increasing older workers' work ability as well as helping organisations to reach their goals. They also pointed out the limitations in applying such an approach, which arise mainly from setting arbitrary age criteria on the basis of either chronological or functional age, and from disregarding differences in the expectations of older workers. To be effective, age management should be proactive, focus on prevention not just on older workers and apply a life-course approach. There is also need to replace randomly used "age-friendly" HR practices and policies with more integrated and strategically relevant HR practices, which allow the fostering of a work environment friendly for all workers regardless of their age (Winkelmann-Gleed 2012). Age management can be treated as the management of organisation's productivity and human resources in a way that acknowledges workers' resources during their individual life course and leads to combating age barriers and promoting age diversity (Wallin and Hussi 2011). Ilmarinen (2006) claims that age management requires taking the given worker's age and age-related factors into account in daily work management, work planning and work organisation in a way that enables everyone, regardless of age, to achieve personal and organisational targets healthily and safely. Thus, age management can benefit both individuals and the organisation and is closely related to health and safety issues. According to Ilmarinen and Rantanen, age management is one of the most effective tools for improving the work ability of older workers (Ilmarinen and Rantanen 1999). Swedish experiences have confirmed that a well-executed age management programme can be a successful method to promote work ability among older workers (Skoglund and Skoglund 2005). This is also confirmed by Bugajska and Makowiec, who emphasise the close relationship between age management safety and health and management (Bugajska et al. 2010).

The overall aim of age management is to limit early exit from the workforce and ensure that working allows healthy workers to maintain their physical and mental health throughout their work–life course, and remain healthy into retirement (Crawford et al. 2016). Since chronological age is not the most appropriate indicator for measuring a person's ability to work, work ability, which takes into account the resources of the individual in relation to the needs of his or her work, can be considered as the best measure for achieving this objective. Work ability is a worker's capacity to do their work with respect to the work demands and their health and mental resources (Ilmarinen and Tuomi 1992). A holistic model of work ability was created by the Finnish Institute of Occupational Health on the basis of studies conducted since the nineties of the occupational well-being of workers of various sectors and age groups. This model is usually presented as "a work ability house" and indicates the relationships between work ability, individual potential and factors related to work and its environment (Figure 5.1).

According to this model, the basis for work ability are individual resources, which include health, physical, mental and social functioning. Such resources affect work ability over the course of one's working life. The next level is knowledge, skills and their continuous improvement, followed by values, attitudes and motivation. The

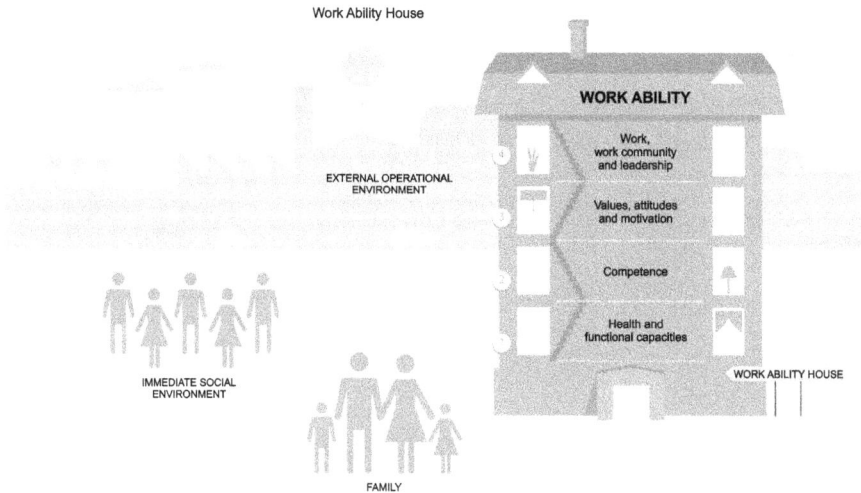

FIGURE 5.1   The workability house model (reprinted with permission from Ilmarinen J., and S. Lehtinen. 2004. Past, Present and Future of Work Ability. People and Work – Research Reports 65)

fourth level covers factors related to work, including material, psychological and social work environment. Factors that foster work ability and affect this "house" include family, the social environment (friends and colleagues), as well as an environment with efficient health care systems which promote health and safety protection. Such a model presents, on one hand, the complexity of processes that have an impact on work ability, and, on the other hand, an indication of the key role that companies have in relation to these issues. Ilmarinen asserts that maintaining work ability requires maintaining a balance between human resources and work. As the preservation of resources is mainly the responsibility of the worker, and work is shaped by the employer, good work ability can only be achieved by co-operation between management and workers. Thus one of the ways to achieve this is participation of workers in planning and implementing actions oriented to foster work ability. Furthermore, the promotion of work ability also improves health and quality of life after the end of working life (Ilmarinen 2005). Each of the levels of the "work ability house" model can be reinforced by actions in the framework of age management (Table 5.1).

Factors affecting participation in the labour market are also the subject of research carried out by the European Foundation for the Improvement of Working and Living Conditions (Eurofound). Eurofound's long-standing research into working conditions and quality of life in EU Member States has provided the basis for the development of sustainable work concept. According to this concept, work is sustainable when working and living conditions support people's involvement and staying in the labour market throughout their extended working lives. Such conditions are created

TABLE 5.1

Dimensions of Work Ability and Related Age Management Elements in a Company

Dimensions of work ability		Age management element
WORK	Physical work demands	• Risk management
	Mental strain	• Stress management
	Quality of supervisory support	• Management commitment and worker participation
		• Communication
	Work independence	• Professional growth
	Opportunities to develop	• Professional growth
VALUES	Work enjoyment Work enthusiasm	• Management commitment and worker participation
		• Communication
EXPERTISE	Basic education	• Recruitment
	Vocational training	• Lifelong learning
		• Management commitment
	Skills	• Lifelong learning
		• Management commitment
HEALTH, FUNCTIONAL CAPACITY	Health and functional abilities	• Risk management
		• Health promotion

when it is possible to adapt work to the needs and capabilities of those who work in such a way that they can and want to work up to retirement age, and even longer (Eurofound 2015). Sustainable work is therefore work that supports work ability. Among different factors influencing sustainable work, such as individual characteristics, work-related elements, social norms and the institutional context, working conditions play an important role. They are linked to three sustainable work characteristics: work–life balance, health and well-being, and also to career prospects (Eurofound 2017). A simplified model illustrating the main dimensions of sustainable work identified in the Eurofound studies is shown in the Figure 5.2.

The work ability and sustainable work models point to similar issues and areas of action that can affect work ability of the working population and that fall within the scope of age management. One of these areas is working conditions, including health and safety at work, which can be shaped within the framework of socially responsible management of health and safety at work.

5.3.2 Implementing Age Management Practices in Socially Responsible Companies

Corporate social responsibility is most often presented from the perspective of good practice, which reflects, among other things, the age management approach of companies. The analysis of 464 good practices submitted in 2014 and 2015 by enterprises implementing the principles of social responsibility to the competition

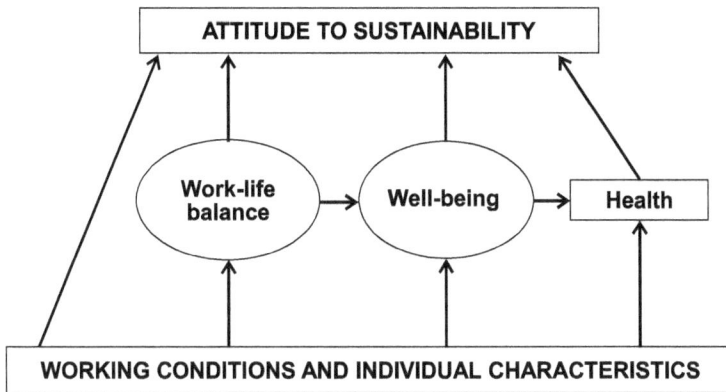

FIGURE 5.2   General structure of the sustainable work model (source: Eurofound [2017])

"Social Reports", organised by the Responsible Business Forum in Poland showed that almost 60% of them are implemented within the framework of human resources management, and far fewer, about 30%, within the framework of OSH management; the smallest number of these practices involve health promotion, which is generally treated as an element of human resources management, but can also be implemented within the framework of occupational health and safety management (Pawłowska and Galwas 2019). Good practices in age management implemented under human resources management involve education and development of workers, improvement of human resources management (including HR strategies and policies, motivation systems, employee opinion surveys, etc.); the smallest number (about 2%) relate directly to diversity issues in the workplace. Among the good practices in the area of OSH management, one can find as the following:

- improvement of OSH management, including improvement of policies, structures and procedures in the field of OSH management, development of monitoring, as well as integration of the OSH management system with other management systems;
- organisational and technical solutions in the field of OSH, including the implementation of various innovative solutions in order to reduce occupational risk;
- safety programmes addressed to selected groups of employees and subcontractors;
- activities promoting safety culture, including safety campaigns, thematic meetings, competitions, magazines and information services, monthly top management visits to production plants and training; and
- working with external partners to promote OSH-related issues.

The practices presented in the social responsibility reports concern broadly understood age management, which is implemented both within the framework of human resources management and OSH management and covers all workers. The age aspect is rarely taken into account directly.

A similar picture is given by a study carried out in 86 enterprises to assess the degree of implementation of age management measures (Pawłowska 2016). The results of this study show that (Figure 5.3):

- age management measures are more often implemented as part of human resources management than OSH management;
- issues related to the age and health of workers are poorly integrated into occupational risk management and monitoring of various aspects of OSH;
- health promotion is one of the least frequent measures implemented to support work ability; and
- the participation of workers in age management is insufficient.

Despite the fact that working conditions, including health and safety at work, have a significant impact on work ability, age management in socially responsible companies is generally considered to be part of human resource management. In the past years very few occupational health-related interventions have addressed the health and work ability of older workers (McDermott et al. 2010; Hildt-Ciupińska and Bugajska 2013). There are also a small number of published documents about occupational health approaches aimed at maintaining the health and work ability of older workers (Poscia et al. 2016). At the same time, studies prove a link between a

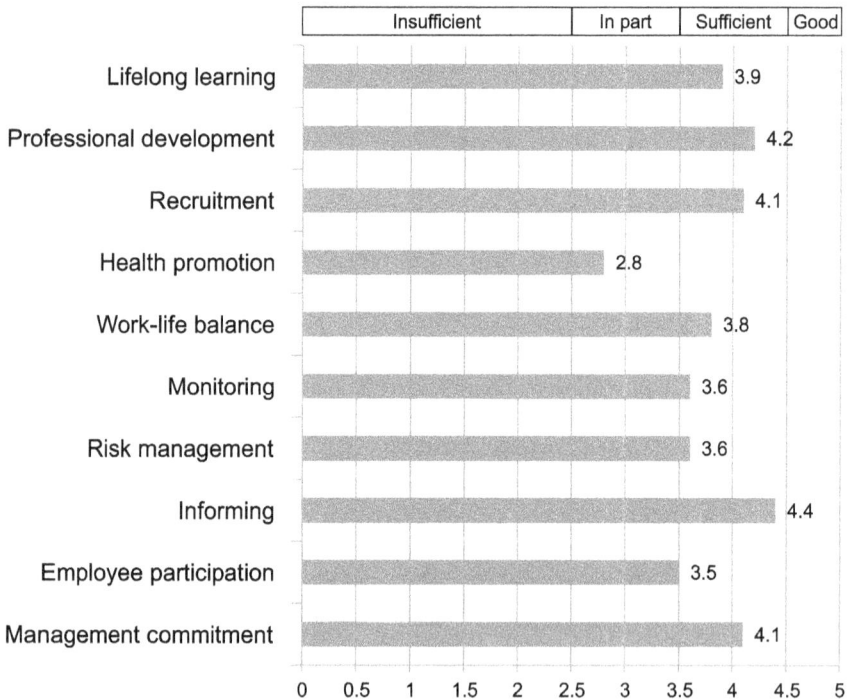

FIGURE 5.3 The level of implementation of various age management elements in companies (average scores for 86 surveyed companies, source: Pawłowska [2016])

proactive approach to health and safety management and improved organisational outcomes, such as improved health and well-being (Ward et al. 2008). This confirms that proactive OSH management may support the promotion of work ability. Close links between age management and occupational safety and health management are also highlighted in EU-OSHA publications (Crawford et al. 2016).

5.3.3 AGE MANAGEMENT AS A COMPONENT OF SOCIALLY RESPONSIBLE OSH MANAGEMENT

Socially responsible OSH management should address issues of social responsibility that are important for ensuring safety, health and well-being of workers. This means, inter alia, that diversity issues, including age diversity, must be taken into account and that elements of age management must be incorporated into OSH management. However, OSH management systems implemented in enterprises often fail to address age-related issues to a satisfactory degree. This has been confirmed by research conducted in 86 companies with and without certified OSH management systems, the results of which demonstrate that in both groups of companies the level of implementation of age management elements is comparable. Results of this research revealed that only some companies develop their OSH management systems in such a way that enables age management to be integrated with OSH management (Pawłowska 2016).

Achieving the basic objective of age management in an enterprise, which is maintaining work ability throughout working life, requires the integration of age management elements into both human resources management and health and safety management (see Table 5.2).

To incorporate age management into OSH management, it is necessary to consider age-related issues when planning, designing and implementing the OSH management system. The standards for OSH management systems developed so far do not sufficiently emphasise diversity issues, including age. The opportunity to incorporate age management in OSH management in a more complete way is provided by the new ISO 45001 standard (ISO 2018). This standard is not prescriptive with regard to the design of the system and enables to integrate different aspects of health and safety, including worker well-being through the organisation's OSH management system. The standard emphasises the need to adapt the scope of OSH management systems to the changing external and internal environment (referred to as the organisational context) and to identify the risks and opportunities arising in this environment.

Demographic changes, and in particular the ageing of the workforce, have an impact on health and safety at work and should be taken into account when analysing the organisational context. The following issues may be relevant when deciding on the inclusion of age management in OSH management:

- the age structure of workers, their needs and expectations;
- the availability of workers with the skills needed for the organisation on the labour market;
- access to information or consultation on effective age management strategies;
- access to medical care;

TABLE 5.2

Basic Elements of Age Management in Human Resources Management and OSH Management

Age management elements in human resources management	Elements of age management in OSH management
Recruitment: • Qualification as a basic criterion for the recruitment of workers • Equal employment opportunities for older and younger workers	**Communication:** • Informing about possible adverse effects of the working environment on their health and about the fact that the harmful effects of certain risks may increase with age • Adapting the methods of communication to the needs of workers
Professional development: • Worker development and career planning, regardless of age	**Risk management:** • Taking into account the age of workers in risk assessment • Introducing changes in the organisation of work, and changing or adapting the workplace where necessary in order to reduce these risks • Adapting the scope and frequency of health surveillance of workers to the risks at work and their individual needs
Lifelong learning: • Providing all workers with training to improve their knowledge and skills for the work they do, if needed • Providing, where necessary, training to facilitate retraining for other work	**Monitoring the work environment:** • Monitoring the psychosocial work environment • Investigating and analysing workers' opinions on risks to their health and safety at work • Collecting and analysing information about the health conditions experienced by workers • Analysing the causes of sick leave of workers
Work–life balance: • Possibility of flexible working time organisation	
Health promotion: • Promoting healthy lifestyle • Providing workers and their families with access to free recreation (swimming pool, gym, etc.) • Facilitating access to health services	**Health promotion:** Actions to reduce work-related stress

Management's involvement in activities aimed at maintaining work ability

Participation of workers in activities aimed at maintaining work ability

---

- cultural determinants and, in particular, stereotypes of older and younger workers;
- differences in the causes of accidents at work and sickness absence of workers in different age groups; and
- the resources needed to plan and implement work ability activities, including human resources.

Analysis of a given organisation's context should lead, inter alia, to the identification of demographic issues which are relevant for the company's performance and related risks and opportunities, and enable early responses.

The ISO 45001 standard also draws attention to developing leadership and management commitment in OSH management as a prerequisite for the success of OSH management systems. Management leadership and involvement is also a prerequisite for integrating age management into OSH management. It is, therefore, important that those responsible for the organisation know and understand age-diversity issues and promote an organisational culture in which diversity is seen as a value to the organisation and which supports the active involvement of all workers in activities supporting work ability. Management's leadership and commitment to health and safety management that takes account of age diversity can be expressed in particular through:

- establishing OSH policy which includes declarations related to maintaining work ability;
- ensuring that, in line with the policy statements, targets and action plans are developed to maintain work ability, taking into account the needs and expectations of workers of different ages and their health status;
- ensuring that OSH management processes (e.g. risk management, communication, monitoring) take account of diversity issues, including age diversity;
- providing the resources needed to incorporate age management into OSH management, including human resources;
- receiving and analysing information on the implementation of activities aimed at maintaining work ability and their results; and
- ensuring the active participation of workers from different age groups in activities aimed at maintaining work ability.

The organisation's OSH policy includes age-diversity issues and may contain, inter alia, statements related to:

- taking account of age diversity in measures to improve working conditions;
- ensuring the development of the qualifications and competences of workers of all age groups;
- supporting reintegration into employment of workers affected by work-related accidents; and
- eliminating work-related stress and counteracting discrimination, harassment and mobbing in the workplace.

Last but not least, worker participation is essential to the effectiveness of OSH management as well as for effective incorporating of age management into OSH management. This may be ensured by:

- informing workers about all matters relating to their safety and health at work, including age-related issues, and, in particular, about possible differences in the physical and intellectual potential of workers in different age groups and the related differences in occupational risks associated with workplace hazards, as well as the possibilities for reducing these risks;

- consulting workers with regard to objectives and plans aimed at maintaining the work ability, and taking into account the views of different age groups when establishing these objectives and plans;
- the participation of workers in decision making on matters relating to the maintenance of work ability; and
- supporting workers' initiatives on measures aimed at maintaining employability in different age groups.

In the requirements of the ISO 45001 standard particular attention is paid to the participation of non-managerial workers in identification of hazards and assessment of occupational risk and actions aimed at its reduction, identification of training needs and evaluation of training, in determining the scope of information needed by workers and the way they are communicated, as well as in investigating incidents and determining corrective actions. The need to consult with regard to their needs and expectations is also emphasised. These requirements support effective mainstreaming of age management into OSH management.

The needs and expectations of workers in relation to the age management measures may be different for groups with different demographic characteristics (Bright 2010; Noone et al. 2016). The study performed in the sample of 93 older workers by Naumanen highlighted that the actions evaluated as the most important for age management include health checks, nursing care, consultancy and training, rehabilitation and psychosocial support. Another survey, conducted among 1,524 people working in 86 enterprises, shows that age management actions implemented under OSH management are, in general, evaluated as more effective than those implemented under human resources management (Figure 5.4). Improving safety and health, promoting healthy behaviour and periodic medical check-ups are considered to be the most effective for maintaining work ability. The periodic medical examinations are more important for respondents in older age groups, while the possibility of obtaining new qualifications and skills or changing jobs is less important for them as for younger workers (Pawłowska 2016).

Age-sensitive risk assessment and management are key to maintaining work ability. In OSH management systems compliant with the ISO 45001 standard the process of hazard identification and risk assessment should take into account the design of work areas, processes, installations, machinery/equipment, operating procedures and work organisation, including the adaptation thereof to the needs and capabilities of the workers involved (ISO 2018). Such a requirement implies, inter alia, the need to incorporate health- and age-related issues into these processes. It should be borne in mind, however, that while the organisations have to adapt environments and jobs to older workers, the effects of biological ageing on individuals are quite heterogeneous (Motta et al. 2005). Age-sensitive risk assessment can be beneficial for all workers, regardless of their age. According to EU-OSHA, key issues to be considered in order to incorporate diversity into the risk assessment process include (EU-OSHA 2009):

- taking diversity issues seriously and having a positive commitment;
- avoiding making prior assumptions about what the hazards are and who is at risk;

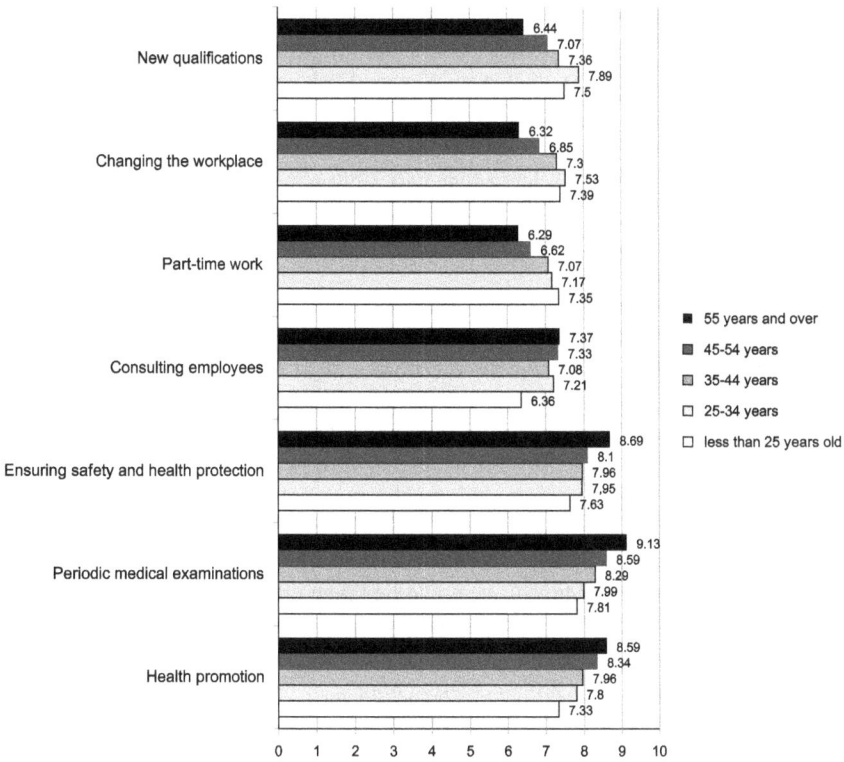

FIGURE 5.4 Evaluations of the importance of different age management measures for workers of different age groups to maintain their work ability, by type of measure (average scores for 1524 surveyed workers, source: Pawłowska [2016])

- valuing the diverse workforce as an asset (and not as a problem);
- considering the entire workforce, including cleaners, receptionists, maintenance workers, temporary agency workers, part-time workers, etc.;
- adapting work and preventive measures to workers;
- considering the needs of the diverse workforce at the design and planning stage;
- linking OSH issues to wider discrimination issues into any workplace equality actions, including equality plans and non-discrimination policies; and
- providing relevant training and information on diversity issues regarding safety and health risks to risk assessors, managers and supervisors, safety representatives, etc.

It is also worth considering the differences in the frequency and severity of accidents between older and younger workers. The analysis of statistical data shows that younger workers are more likely to have accidents at work, but the number of days of absence caused by an accident increases with age (Ordysiński 2013). Additionally, in many publications it is emphasised that age-sensitive risk assessment should be based on a participatory approach (Crawford et al. 2016). The need for a participatory approach to risk assessment is clearly defined in ISO 45001, which requires non-management workers to participate in hazard identification and risk and opportunity assessment, as well as in activities aimed at eliminating hazards and reducing health and safety risks.

To take account of the age of workers in risk assessment, it is feasible to identify types of work and work tasks that may (but need not necessarily) involve an increased level of risk for people of different ages and health status. The following, inter alia, should be taken into account:

- work that involves risks likely to be higher for older workers, in particular work involving heavy physical effort, e.g. manual handling of heavy objects, work in hot or cold microclimates, shift work;
- work that may involve an increased level of risk for young and low-skilled workers, where inexperience or an increased willingness to engage in dangerous behaviour may result in accidents at work;
- work which requires increased psycho-physical fitness to be performed (e.g. work at a height, work requiring high concentration and simultaneous performance of many activities), where the safety of the worker may depend on his or her fitness; it should be taken into account that conditions of reduced fitness which are normal physiological states, such as fatigue or stress occurring in certain circumstances may be more frequent in people with chronic diseases and in elderly people.

Where such work is identified, additional measures may be necessary for those persons or groups of persons for whom the occupational risks associated with the work may be increased. This may include in particular:

- changes to workstation equipment (e.g. change of tools, introduction of additional equipment to eliminate lifting, physical exertion, etc.) in order to adapt it to the worker's capabilities;
- adapting the organisation of work to the needs of the worker (e.g. introducing additional breaks, allowing work to be done at one's own pace);
- entrusting work to teams composed of workers of different ages, performing tasks adapted to their psycho-physical condition;
- the introduction of the principle that work relating to increased occupational risks for young workers should be carried out under the supervision of a mentor;
- the introduction of an obligation for "dynamic risk assessment", as a result of which the supervisor may remove an worker from work when his or her psycho-physical condition does not ensure safe performance of the work; and
- adjusting the scope and frequency of medical examinations of workers to the risks at work and their individual needs, taking into account age.

New opportunities for managing occupational risks, taking into account diversity issues, offer the option of using intelligent personal protective equipment and big data analytics methods, which make possible the real-time monitoring of the health and well-being of the workers and the appropriate adaptation of protective measures (Podgórski et al. 2016).

The requirements of ISO 45001 indicate the need for socially responsible management of occupational safety and health, which takes into account the issues of diversity, including age-related diversity. The scope of this management is not limited to the traditional understanding of safety and health at work, and increasingly covers issues related to improving the health and well-being of workers that are relevant to their work ability. Such OSH management includes elements of age management and must be implemented in close integration with human resources management.

5.4 CHOOSING INDICATORS FOR EVALUATION OF SOCIALLY RESPONSIBLE OSH MANAGEMENT

5.4.1 CSR AND OSH PERFORMANCE MEASUREMENT

Measuring CSR has for years been considered to be an important and difficult task. The basis of CSR measurement is often analysis of the contents of companies' annual reports, expert evaluations, single- and multiple-issue indicators and/or surveys of managers (Giannarakis et al. 2010). Scales measuring CSR at the individual and at the organisational levels have also been proposed (Turker 2009). To ensure the comparable measures, different indexes and tools for the CSR measurement and benchmarking have been developed by different institutions. According to Hopkins (2005) the most commonly used ones include:

- The Corporate Responsibility Index, which is a management and benchmarking tool developed by the UK's Business in the Community (BITC); this is a voluntary, business-led benchmarking index which provides information on the four main impact sections of corporate responsibility: community, workplace, marketplace, and environment (El-Masry and Kamal 2013).
- The FTSE4Good Index Series, which is maintained by the FTSE Group (Financial Times Stock Exchange) to measure the performance of companies that meet globally recognised corporate responsibility standards; to be included in the FTSE4Good Indexes, companies must, inter alia, support human rights, have good relationships with various stakeholders, make progress to become environmentally sustainable, ensure good labour standards (not only for their own company but for companies that supply them as well), and fight bribery and corruption (FTSE Russell 2019).
- The Dow Jones Sustainability Indexes (DJSI), developed to track the performance of the companies leading in terms of corporate sustainability, the Dow Jones Sustainability Indexes consist of a global, European, North American, Asia Pacific and Korean set of indexes. The leading companies are identified in different industry groups on the basis of a systematic assessment and for each group app. 50 general and industry-specific criteria are used to assess economic, environmental and social performance. The criteria are related to, inter, to labour practices, including occupational health and safety. The methodology of assessment is continuously updated (Wikipedia 2019).

- Business Ethics 100, which is the ranking of US companies according to service to seven stakeholder groups (Clarke and Gibson-Sweet 2012).
- The AccountAbility (AA) Rating, a global index developed by the British NGO Accountability in co-operation with training company CSR Network. which is used to measure the impact of the 100 largest international corporations on the social and environmental environment. The ranking is a tool that shows not only whether a company meets the basic standards of responsible business but also whether responsible business is an integral part of the company's strategy (AA1000 Accountability Principles Standard 2008).
- Global Reporting Initiative (GRI) Sustainability Reporting Guidance, which is intended to be used as a framework for voluntary reporting on economic, environmental, and social performance of an organisation. Social performance indicators are proposed in the guidance for different areas of companies impacts. Nowadays indicators for occupational health and safety have been proposed in the GRI 403 standard (GRI 403 2018).

When analysing and comparing CSR measurement systems listed above, Hopkins takes into account the following aspects: concept used and definition; conceptual framework; methodology; indicators and measures proposed; and data available. His analysis shows that only two of the systems considered (GRI and Business Ethics 100) define indicators and measures which can be used as a basis for CSR evaluation and that the conceptual framework which is the basis of choosing components and indicators for measurement is in numerous cases unclear (Hopkins 2005). Another issue is that measurements systems which aim to cover social and environmental aspects by providing a large number of interconnected performance indicators do not provide information on how they are linked to the company's strategy and are not sufficient to identify cause and effect relationships (Panayiotou et al. 2009). Despite these limitations, the CSR measurement principles proposed by different systems may support companies in disclose high quality, relevant, useful, consistent and more comparable non-financial (environmental, social and governance-related) information in a transparent way.

The improvement of existing CSR measurement systems makes them more and more adapted to the changing circumstances and needs related to the evaluation and reporting of activities in various areas of social responsibility. A good example is the GRI 403 Occupational Health and Safety standard developed in 2018, which sets out reporting requirements on the topic of occupational health and safety. This standard is in line with ISO 45001 standard and addresses the need to ensure the safety and health protection of all workers, including those whose work and/or workplace is controlled by the given organisation. It provides requirements and guidance on which information should be presented in companies' sustainability reports including management approach and OSH-specific information. Requirements for information characterising management approach are consistent with those set for OSH management systems in the ISO 45001 standard and relate to, in particular, the implementation of the management system, risk assessment and incident investigation, health care, worker participation and training (Table 5.3). The standard

TABLE 5.3

Basic Information on Management Approach Reported According to GRI 403 Standard

Scope of reporting	Reported information
Occupational health and safety management system	• A statement of whether an OSH management system has been implemented • The basis of the system implementation (legal requirements or recognised risk management and/or management system standards/guidelines) • The scope of occupational health and safety management system
Hazard identification, risk assessment, and incident investigation (information for employees and for workers who are not employees but whose work and/or workplace is controlled by the organisation)	• The processes used to identify work-related hazards and assess risks on a routine and non-routine basis, and to apply the hierarchy of controls in order to eliminate hazards and minimise risks • Information on how the organisation ensures the quality of these processes, including the competency of persons who carry them out and how the results of these processes are used to evaluate and continually improve the occupational health and safety management system • The processes for workers to report work-related hazards and hazardous situations, and an explanation of how workers are protected against reprisals • The policies and processes for workers to remove themselves from work situations that they believe could cause injury or ill health, and an explanation of how workers are protected against reprisals • The processes used to investigate work-related incidents, including the processes to identify hazards and assess risks relating to the incidents, to determine corrective actions using the hierarchy of controls, and to determine improvements needed in the occupational health and safety management system
Occupational health services	• The functions of occupational health services' that contribute to the identification and elimination of hazards and minimisation of risks • An explanation of how the organisation ensures the quality of these services and facilitates workers' access to them
Worker participation, consultation, and communication on occupational health and safety	• The processes for worker participation and consultation in the development, implementation, and evaluation of the occupational health and safety management system, and for providing access to and communicating relevant information on occupational health and safety to workers • Where formal joint management–worker health and safety committees exist, a description of their responsibilities, meeting frequency, decision-making authority, and whether and, if so, why any workers are not represented by these committees
Worker training on occupational health and safety	• Occupational health and safety training provided to workers, including generic training as well as training on specific work-related hazards, hazardous activities, or hazardous situations

(Continued)

TABLE 5.3 (CONTINUED)

Basic Information on Management Approach Reported According to GRI 403 Standard

Scope of reporting	Reported information
Promotion of worker health	• An explanation of how the organisation facilitates workers' access to non-occupational medical and healthcare services, and the scope of access provided • Any voluntary health promotion services and programmes offered to workers to address major non-work-related health risks, including the specific health risks addressed, and how the organisation facilitates workers' access to these services and programmes
Prevention and mitigation of occupational health and safety impacts directly linked by business relationships	• A description of the organisation's approach to preventing or mitigating significant negative occupational health and safety impacts that are directly linked to its operations, products or services by its business relationships, and the related hazards and risks

emphasises the need to use and include in the reports leading indicators used to measure an organisation's OSH performance. It also defines basic indicators for evaluating OSH outputs (Table 5.4).

5.4.2 Lagging Indicators for Monitoring Socially Responsible OSH Management

Lagging indicators serve to assess results of actions performed. OSH activities are traditionally evaluated using the indicators relating to accidents at work and occupational diseases. Such indicators are also recommended to for assessment of OSH performance in the GRI 403 guidelines. The most commonly used, and one of the simplest lagging indicators, is the number of accidents at work (in which the number of total, fatal and serious accidents is usually presented separately) per period of time (e.g. month, quarter, year). This indicator may be used to monitor the number of accidents over time and can be applied to a specific group of workers, e.g. all employees of a company, employees of selected departments, age groups, etc. However, it is not useful for comparisons between groups of different sizes and different companies. To enable such comparisons, occupational accident frequency and severity rates may be used. Frequency rates can be calculated for all accidents that have occurred or separately for accidents with different consequences (e.g. non-fatal and fatal). The accident rate may be calculated as the number of accidents per 100,000 employees (full-time equivalents) or as the number of accidents at work per defined number of hours worked, usually 100,000 or 200,000 man-hours. Severity rates are used to characterise the severity of the accidents, which can be done by referring the total number of days of absence caused by accidents at work in a specific time (e.g. quarter, year) to the number of people injured in these accidents.

TABLE 5.4

Basic OSH-Specific Information Reported According to GRI 403 Standard

OSH issue	Indicators
Workers covered by an occupational health and safety management system (if the organisation has implemented an occupational health and safety management system based on legal requirements and/or recognised standards/guidelines)	• The number and percentage of all employees and workers who are not employees but whose work and/or workplace is controlled by the organisation, who are covered by a system (system may be not audited, internally audited or audited or certified by an external party) • Workers excluded from this disclosure, including the types of worker excluded
Work-related injuries (for all employees and for all workers who are not employees but whose work and/or workplace is controlled by the organisation)	• The number and rate of fatalities as a result of work-related injury • High-consequence work-related injuries (excluding fatalities) and recordable work-related injuries; the main types of work-related injury and the number of hours worked • The work-related hazards that pose a risk of high-consequence injury, and actions taken or underway to eliminate these hazards and minimise risks using the hierarchy of controls • Any actions taken or underway to eliminate other work-related hazards and minimise risks using the hierarchy of controls
Work-related ill health (for all employees and for all workers who are not employees but whose work and/or workplace is controlled by the organisation)	• The number of fatalities as a result of work-related ill health, the number of cases of recordable work-related ill health and the main types of work-related ill health • The work-related hazards that pose a risk of ill health, and actions taken or underway to eliminate these hazards and minimise risks using the hierarchy of controls

Socially responsible OSH management increasingly covers workers' health issues. The lagging indicators used to assess the effectiveness of actions aimed at protecting workers' health are mainly the number of work-related ill health (diseases, illnesses and disorders) or work-related absence. The number of days of sick leave for workers is in most cases known; it is more difficult to obtain the data needed to determine the number of days of work-related absence. However, it is possible to obtain information on work-related risks and the number of people exposed to them. Such an indicator may be even more useful for assessing the effectiveness of action on the prevention of work-related illnesses because, in many cases, the adverse health effects of risks are delayed.

To evaluate the performance of socially responsible management system, additional indicators may be also used with the aim of measuring the work sustainability in a company. According to the model developed by Eurofound, indicators such as subjective employee assessments of health, mental well-being (welfare), work–life balance and job insecurity and employability can be used to measure the level of work sustainability. Since these indicators are subjective in nature, it is necessary to conduct surveys among employees in order to collect the information needed to calculate them. The use of such indicators can give a more complete picture of an organisation's performance in terms of its contribution to achieving the sustainable development goals. However, studies show that subjective indicators (e.g. subjective assessments of working conditions), which require employee surveys, are the least frequently used in companies (Pawłowska 2015).

5.4.3 LEADING INDICATORS FOR MONITORING ACTIONS IN THE FRAMEWORK OF SOCIALLY RESPONSIBLE OSH MANAGEMENT

Leading indicators are used to measure companies' actions that are expected to improve social, environmental and economic performance. They complement the traditional lagging indicators and provide the information necessary for monitoring and improving implemented actions taken into account the changing circumstances. The relationship between some leading and lagging indicators has been confirmed by research (Pawłowska 2015; Pęciłło 2012). The need to use leading indicators to evaluate processes in the OSH management system is emphasised in the ISO 45001 standard as well as the GRI 403 guidelines. Which leading indicators will be used depends on the needs of the company, and particularly on the priorities and action plans adopted. Examples of leading indicators for monitoring and assessment of actions related to different issues of social responsibility in the framework of OSH management system are presented in Table 5.5.

Leading indicators are, in general, applied in companies less frequently than lagging indicators. According to the results of the survey performed in 60 companies, the most frequently leading indicators are used to evaluate actions required by law, such as OSH trainings and risk assessments. The number of leading indicators used and regularly monitored increases with the level of implementation of OSH management practices in accordance with voluntary standards (Pawłowska 2015). It can therefore be assumed that legal requirements, as well as occupational health and safety management standards (such as ISO 45001) and social responsibility reporting standards (such as GRI 403), significantly support the use of leading indicators for monitoring occupational health and safety in enterprises.

5.5 SUMMARY

The response to the challenges posed by changes in the world of work, including the increasing digitisation and automation of work processes, demographic change and new forms of employment, demands a new approach to OSH management that aims not only at preventing work-related injuries and illnesses, but also at improving the health and well-being of workers, thereby prolonging their working lives. The new approach means that OSH management covers an increasing number of issues relevant to corporate social responsibility and is increasingly integrated into the overall management of an organisation. By adopting such an approach, organisations manage occupational safety and health in a socially responsible manner, contributing to the achievement of sustainable development objectives, and in particular Objective 8

TABLE 5.5
Leading Indicators for Monitoring and Evaluating Actions Targeted at Different Issues of Social Responsibility in the Framework of OSH Management Systems

Type of action	Leading indicator
Examples of actions targeted at vulnerable groups	
Adapting workplaces to the needs of workers in terms of their state of health and physical and mental capacity	Number of adapted workplaces per year Number of adapted workplaces in relation to the number of workplaces for which the need was identified
Programmes/measures aimed at retaining older workers in employment	Number of workers covered by the programme
Programmes/measures aimed at ensuring health protection of younger workers	Number of workers covered by the programme in relation to the total number of workers in a given category
Programmes/actions targeted at other vulnerable groups (e.g. people with disabilities, migrant workers, etc.)	
Examples of actions targeted at eliminating violence at work and improving the psychosocial working environment	
Management training on stress and how to deal with it	Average number of training hours per person
Employee training on psychosocial risks (stress, mobbing, bullying)	Average number of training hours per person
Work-related stress reduction programmes	Number of workers covered by the programme Number of workers covered by the programme in relation to the total number of workers
Examples of actions targeted at safety and health at work	
Identifying and controlling safety and health risks associated with the company's activities	Number of hazards identified in quarter/year Number of hazards eliminated in the quarter/year Number of new (better) protection measures introduced in the quarter/year
Recording and analysing accidents at work, diseases and health problems reported by employees	Periodic reports on work-related sickness absenteeism Periodic surveys of work-related health problems Periodic accident investigation reports
Implementation of ergonomic improvements, including adaptation of workstations to the needs of employees	Number of ergonomic improvements implemented per year Number of employees submitting proposals for ergonomic improvements
Examples of actions targeted at health promotion at work and rehabilitation	
Monitoring sick leave of employees	Periodic reports of sickness absenteeism examinations
Providing rehabilitation for workers who have suffered an accident or work-related illness	Number of workers who were provided with rehabilitation

(Continued)

TABLE 5.5 (CONTINUED)

Leading Indicators for Monitoring and Evaluating Actions Targeted at Different Issues of Social Responsibility in the Framework of OSH Management Systems

Type of action	Leading indicator
Providing training for workers involved in accidents to enable them to retrain	Number of workers who received training
Implementation of programmes aimed at promoting a healthy lifestyle	Number of persons covered by the programme
Ensuring access to free forms of recreation for workers and their families	Number of persons provided with access to free forms of recreation
Ensuring access to health services for and their families	Number of people provided with access to medical services
Examples of actions aimed at human development	
Vocational training	Number of people provided with training
Training in occupational risk assessment and health and safety management	Number of persons who were provided with training in relation to the number of persons who needed it
Periodic evaluation of employees, taking into account their attitudes and commitment in safety and health protection	Percentage of employees receiving performance reviews and information on career development

which is to promote sustained, inclusive and sustainable economic growth, full and productive employment and decent work for all.

Socially responsible OSH management requires the developing of effective governance within OSH management systems and incorporating into those systems CSR issues related to OSH. This means, in particular, that OSH management systems should take account of problems in the psychosocial working environment, as well as diversity issues.

Mainstreaming age diversity in OSH management involves integrating age management elements into OSH management processes. The key role is played by age-sensitive risk assessment and the resulting actions, which in some cases (such as the need to change the workplace, retrain, change the organisation of working time) should be introduced in close co-ordination with human resources management. Health promotion is also becoming more and more important and can be integrated into both OSH and human resources management.

To integrate effectively social responsibility in OSH management, it is also necessary to monitor and measure chosen aspects of OSH performance. Measurement and evaluation of the actions undertaken and progress made in achievement of established objectives may be done using different indicators (leading for evaluating actions and lagging for achievement of objectives).

Socially responsible OSH management is expected to respond effectively to the new challenges in working life. Its implementation involves developing the competences and capabilities of management and employees in a way that enables them to understand social responsibility issues related to safety and health and to commit to socially responsible OSH management.

REFERENCES

AA1000 AccountAbility Principles Standard. 2008. https://www.accountability.org/wp-c ontent/uploads/2018/05/AA1000APS-2008.pdf (accessed September 18, 2019).

Aaltio, I., H. Salminen, and S. Koponen. 2011. Ageing employees and human resource management evidence of gender-sensitivity? *Equality, Diversity and Inclusion: Anais an International Journal* 33(2):160–176.

Badri, A., B. Boudreau-Trudel, and A. Saâdeddine Souissi. 2018. Occupational Health and safety in the industry 4.0 era: A cause for major concern? *Safety Science* 109:403–411.

Bright, L. 2010. Why age matters in the work preferences of public employees: A comparison of three age-related explanations. *Public Personnel Management* 39(1):1–14.

Bugajska, J., T. Makowiec-Dąbrowska, and E. Wągrowska-Koski. 2010. Age management in enterprises as a part of occupational safety and health in elderly workers. *Medical Section Proceedings* 61(1):55–63. http://www.imp.lodz.pl/upload/oficyna/artykuly/pdf /full/2010/2010_3/2010_1/MP_1-2010_J-Bugajska.pdf (accessed September 18, 2019).

Clark, A. E. 1998. Measures of job satisfaction: What makes a good job? Evidence from OECD countries, OECD Labour Market and Social Policy Occasional Papers 34. Paris: OECD Publishing. https://ideas.repec.org/p/oec/elsaaa/34-en.html (accessed September 18, 2019).

Clarke, J., and M. Gibson-Sweet. 2012. The use of corporate social disclosures in the management of reputation and legitimacy: A cross sectoral analysis of UK top 100 companies. *Business Ethics: A European Review.* https://onlinelibrary.wiley.com/doi/pdf/10.11 11/1467-8608.00120 (accessed September 18, 2019).

Crawford, J., A. Davis, H. Cowie, and K. Dixon. 2016. The ageing workforce: Implications for OSH – A research review. European Agency for Safety and Health at Work. https:// osha.europa.eu/en/publications/ageing-workforce-implications-occupational-safety-a nd-health-research-review/view (accessed September 18, 2019).

Dahlsrud, A. 2006. How corporate social responsibility is defined: An analysis of 37 definitions. *Corporate Social Responsibility and Environmental Management* 15(1):1–13.

Dekker, W. A. 2014. The bureaucratization of safety. *Safety Science* 70:348–357. https:// www.safetydifferently.com/wp-content/uploads/2014/08/BureaucratizationSafety.pdf (accessed September 18, 2019).

El-Masry, A., and N. Kamal. 2013. Corporate responsibility index. In: Idowu S. O., Capaldi N., Zu L., Gupta A. D. (eds.) *Encyclopedia of Corporate Social Responsibility.* Berlin, Heidelberg: Springer.

EU-OSHA OSHA (European Agency for Safety and Health at Work). 2009. Workforce diversity and risk assessment – Ensuring everyone is covered. European Agency for Safety and Health at Work. https://osha.europa.eu/en/publications/reports/TE7809894ENC/ view (accessed September 18, 2019).

Eurofound (European Foundation for the Improvement of Living and Working Conditions). 2015. Sustainable work over the life course: Concept paper. Luxembourg: Publications Office of the European Union. https://www.eurofound.europa.eu/pl/publications/re port/2015/working conditions/sustainable-work-over-the-life-course-concept-paper (accessed September 18, 2019).

Eurofound (European Foundation for the Improvement of Living and Working Conditions). 2017. Working conditions of workers of different ages: European Working Conditions Survey 2015. Luxembourg: Publications Office of the European Union. https://ww w.eurofound.europa.eu/publications/report/2017/working-conditions-of-workers-of-different-ages (accessed September 18, 2019).

Frick, K. 2011. Worker influence on voluntary OSH management systems: A review of its ends and means. *Safety Science* 49(7):974–987.

FTSE Russell. FTSE4Good index series. https://www.ftserussell.com/products/indices/ftse4 good (accessed September 18, 2019).

Gallagher, C., and E. Underhill. 2012. Managing work health and safety: Recent developments and future directions. *Asia Pacific Journal of Human Resources* 50(2):227–244.

Gallagher, C., E. Underhill, and M. Rimmer. 2003. Occupational safety and health management systems in Australia: Barriers to success. *Policy and Practice in Health and Safety* 1(2):67–81.

Giannarakis, G., D. Galani, G. Charitoudi, and N. Litinas. 2010. The weight of corporate social responsibility indicators in measurement procedure, world academy of science, engineering and technology. *International Journal of Social, Behavioral, Educational, Economic, Business and Industrial Engineering* 4(6):409–417. https://pdfs.semanticscholar.org/7d 0a/37b24de223e5113d7ae627faefe3dd55f65f.pdf (accessed September 18, 2019).

Global Reporting Initiative (GRI), 2018. GRI 403: Occupational health and safety, GRI, Amstrdam, The Netherlandes. https://www.globalreporting.org/standards/media/1910/gri-403-occupational-health-and-safety-2018.pdf (accessed September 18, 2019).

Hale, A., F. Guldenmund, P. van Loenhout, and J. Oh. 2010. Evaluating safety management and culture interventions to improve safety: Effective intervention strategies. *Safety Science* 48(8):1026–1035.

Hasle, P., and G. Zwetsloot. 2011. Editorial: Occupational health and safety management systems: Issues and challenges. *Safety Sci* 49(7):961–963.

Hildt-Ciupińska, K., and J. Bugajska. 2013. Evaluation of activities and needs of older workers in the context of maintaining their employment. *Medical Section Proceedings* 64(3):297–306. http://www.imp.lodz.pl/upload/oficyna/artykuly/pdf/full/2013/%20--2013_3_hildtciupinska.pdf (accessed September 18, 2019).

Hohnen, P., P. Hasle, A. Jespersen, and C. Madsen. 2014. Hard work in soft regulation: A discussion of the social mechanisms in OHS management standards and possible dilemmas in the regulation of psychosocial work environment. *Nordic Journal of Working Life Studies* 4(3):13–30. https://tidsskrift.dk/njwls/article/view/26716/23488 (accessed September 18, 2019).

Hopkins, M. 2005. Measurement of corporate social responsibility. *International Journal of Management and Decision Making* 6(3–4):213–231.

Ilmarinen, J. 2005. Towards a longer worklife – Ageing and the quality of worklife in the European Union. Helsinki: Finnish Institute of Occupational Health. http://hawai4u.de/UserFiles/Ilmarinen_2005_Towards%20a%20Longer%20Worklife.pdf (accessed September 18, 2019).

Ilmarinen, J. 2006. The ageing workforce – Challenges for occupational health. *Occupational Medicine* 56(6):362–364. http://occmed.oxfordjournals.org/content/56/6/362.short (accessed September 18, 2019).

Ilmarinen, J., and J. Rantanen. 1999. Promotion of work ability during ageing. *American Journal of Industrial Medicine* (Supplement 1):21–23. https://onlinelibrary.wiley.com/doi/epdf/10.1002/%28SICI%291097-0274%28199909%2936%3A1%2B%3C21%3A%3AAID-AJIM8%3E3.0.CO%3B2-S (accessed September 18, 2019).

Ilmarinen, J., and K. Tuomi. 1992. Work ability of aging workers. *Scandinavian Journal of Work, Environment and Health* 18(Suppl 2):8–10. http://www.sjweh.fi/show:abstract.php?abstract_id=1627 (accessed September 18, 2019).

ISO (International Organization for Standardization). 2010. *ISO 26000: Guidance to Social Responsibility*. Geneva, Switzerland: International Organization for Standardization.

ISO (International Organization for Standardization). 2018. *ISO 45001:2018, Occupational Health and Safety Management Systems – Requirements With Guidance for Use.* Geneva, Switzerland: International Organization for Standardization.

Jain, A., S. Leka, and G. Zwetsloot. 2018. Managing health, safety and wellbeing, – ethics, responsibility and sustainability, Springer Netherlands. https://link.springer.com/book /10.1007/978-94-024-1261-1 (accessed September 18, 2019).

McDermott, H. J., A. Kazi, F. Munir, and C. Haslam. 2010. Developing occupational health services for active age management. *Occupational Medicine* 60(3):193–204. https:// pdfs.semanticscholar.org/d68b/156fca47b7e45e8f7678f3886b8acc9e67f4.pdf (accessed September 18, 2019).

Micheal, B. 2003. Corporate social responsibility in international development: An overview and critique. *Corporate Social Responsibility and Environmental Management* 10(3):115–128. http://www.ceads.org.ar/downloads/Making%20good%20business%2 0sense.pdf (accessed September 18, 2019).

Moir, L. 2001. What do we mean by corporate social responsibility? *Corporate Governance: The International Journal of Business in Society* 1(2):16–22. https://core.ac.uk/downlo ad/pdf/138652.pdf (accessed September 18, 2019).

Montero, M. J., R. A. Araque, and J. M. Rey. 2009. Occupational health and safety in the framework of corporate social responsibility. *Safety Science* 47(10):1440–1445.

Motta, E., L. Bennati, M. Ferlito, and L. Alaguamera 2005. Successful aging in centenarians: Myths and reality. *Archives of Gerontology and Geriatrics* 40(3):241–251. https://ww w.sciencedirect.com/science/article/pii/S0167494304001670?via%3Dihub (accessed September 18, 2019).

Mullerat, R. 2011. *International Corporate Social Responsibility: The Role of Corporations in the Economic Order of the 21st Century*, Kindle Edition. Kluwer Law International, the Netherlands

Naegele, G., and A. Walker. 2006. A guide to good practice in age management. European Foundation for the Improvement of Living and Working Conditions. http://eurofoun d.europa.eu/sites/default/files/ef_files/pubdocs/2005/137/en/1/ef05137en.pdf (accessed September 18, 2019).

Noone, J., A. Knox, K. O'Loughlin, M. McNamara, P. Bohle, and M. Mackey. 2016. An analysis of factors associated with older workers' employment participation and preferences in Australia. *Fronties in Psychology* 9:2524. Published online 2018 Dec 19; 9:2524.

Ordysiński, S. 2013. Częstość wypadków przy pracy a wiek i staż poszkodowanych. *Bezpieczeństwo Pracy – Nauka i Praktyka* 7:22–26.

Panayiotou, N. A., K. G. Aravossis, and P. Moschou. 2009. A new methodology approach for measuring corporate social responsibility performance. *Water, Air, and Soil Pollution: Focus* 9(1–2):129–138. https://link.springer.com/article/10.1007/s11267-008-9204-8 (accessed September 18, 2019).

Pawłowska, Z. 2011. Wdrażanie zasad odpowiedzialności społecznej w systemach zarządzania bezpieczeństwem i higieną pracy a jakość życia w pracy (Implementing social responsibility in OSH management and quality of working life). *Bezpieczeństwo Pracy – Nauka i Praktyka* 4(475):16–18.

Pawłowska, Z. 2013. Occupational safety and health management and corporate social responsibility (CSR), OSH-WIKI-article. https://oshwiki.eu/wiki/Occupational_safety_and_health_ management_and_corporate_social_responsibility_(CSR) (accessed September 18, 2019).

Pawłowska, Z. 2015. Using lagging and leading indicators for the evaluation of occupational safety and health performance in industry. *International Journal of Occupational Safety and Ergonomics: JOSE* 21(3):284–290.

Pawłowska, Z. 2016. Integrowanie zarządzania wiekiem z zarządzaniem zasobami ludzkimi i bezpieczeństwem i higieną pracy (Integrating age management with OSH management and HR management). *Problemy Jakości* 4:17–21.

Pawłowska, Z., and M. Galwas. 2019. Wytyczne do działań na rzecz utrzymywania zdolności do pracy w ramach wdrażania koncepcji społecznej odpowiedzialności przedsiębiorstw (CSR). *CIOP-PIB Research Report* (Unpublished).

Pęciłło, M. 2012. Results of implementing programmes for modifying unsafe behaviour in Polish companies. *International Journal of Occupational Safety and Ergonomics: JOSE* 18(4):473–485.

Podgórski, D. 2005. Workers' involvement – A missing component in the implementation of occupational safety and health management systems in enterprises. *International Journal of Occupational Safety and Ergonomics: JOSE* 11(3):219–231.

Podgórski, D., K. Majchrzycka, A. Dąbrowska, G. Gralewicz, and M. Okrasa. 2016. Towards a conceptual framework of OSH risk management in smart working environments based on smart PPE, ambient intelligence and the Internet of Things technologies. *International Journal of Occupational Safety and Ergonomics: JOSE* 23(1):1–20.

Poscia, A., U. Moscato, D. La Milia, et al. 2016. Workplace health promotion for older workers: A systematic literature review. *BMC Health Servey Research* 16(Suppl 5):329. https://www.ncbi.nlm.nih.gov/pmc/articles/PMC5016729/ (accessed September 18, 2019).

Robson, L., J. Clarke, K. Cullen, et al. 2007. The effectiveness of occupational health and safety management systems: A systematic review. *Safety Science* 45(3):329–353.

Rocha, R. 2010. Institutional effects on occupational health and safety management systems. *Human Factors and Ergonomics in Manufacturing and Service Industries* 20(3):211–225.

Skoglund, B., and C. Skoglund. 2005. Can age management promote work ability among older workers? *International Congress Series* 1280:392–396. http://www.sciencedirect.com/science/journal/05315131/1280 (accessed September 18, 2019).

Sowden, P., and S. Sinha. 2005. Promoting health and safety as a key goal of the Corporate Social Responsibility agenda, HSE. http://www.hse.gov.uk/research/rrpdf/rr339.pdf (accessed September 18, 2019).

TAEN (The age and Employment Network). 2007. Managing the ageing workforce: An introductory guide to age management for HR professionals. http://www.ageingatwork.eu/resources/taen-guide-to-age-management-sept-07.pdf (accessed September 18, 2019).

Turker, D. 2009. Measuring corporate social responsibility: A scale development study. *Journal of Business Ethics* 85(4):411–427.

United Nations. 2015. Transforming our world: The 2030 agenda for sustainable development. https://sustainabledevelopment.un.org/content/documents/21252030%20Agenda%20for%20Sustainable%20Development%20web.pdf (accessed September 18, 2019).

Walker, A. 1997. Combating age barriers in employment – A European Research Report, European Foundation, and Dublin. https://www.eurofound.europa.eu/publications/report/2005/labour-market/combating-age-barriers-in-employment-european-research-report (accessed September 18, 2019).

Walker, A. 1998. Managing an ageing workforce: A guide to good practice. Dublin: European Foundation for the Improvement of Living and Working Conditions. http://edz.bib.uni-mannheim.de/www-edz/pdf/ef/98/ef9865en.pdf (accessed September 18, 2019).

Wallin, M., and T. Hussi. 2011. Best practices in age management – Evaluation of organisation cases, Final report. Finnish Institute of Occupational Health. https://www.tsr.fi/c/document_library/get_file?folderId=13109&name=DLFE-8009.pdf (accessed September 18, 2019).

Ward, J., C. Haslam, and R. Haslam. 2008. The impact of health and safety management on organisations and their staff. Institute of Occupational Safety and Health, Research Report 08.1. Leicester: IOSH. https://www.iosh.com/resources-and-research/resources/impact-health-safety-management-on-organisations-their-staff/ (accessed September 18, 2019).

WHO (World Health Organization). 1994. Global strategy on occupational health for all: The way to health at work, recommendation of the second meeting of the WHO Collaborating Centres in Occupational Health, 11–14 October 1994, Beijing, China. https://www.who.int/occupational_health/publications/globstrategy/en/index5.html (accessed September 18, 2019).

Wikipedia, Dow Jones Sustainability Indices. https://en.wikipedia.org/wiki/Dow:Jones_
Sustainability_Indices (accessed September 18, 2019).

Winkelmann-Gleed, A. 2012. Retirement or committed to work? Conceptualising prolonged
labour market participation through organizational commitment. *Employee Relations*
34(1):80–90.   http://www.emeraldinsight.com/doi/abs/10.1108/01425451211183273?j
ournalCode=er (accessed September 18, 2019).

Zwetsloot, G. 2003. From management systems to corporate social responsibility. *Journal of
Business Ethics* (Special issue on Corporate Social Responsibility) 44:201–207.

Zwetsloot, G., and A. Starren. 2004. *Corporate Social Responsibility and Safety and Health
at Work, European Agency for Safety and Health at Work.* Luxembourg: Office for
Official Publications of the European Communities.

6 Enhancing OSH Management Processes through the Use of Smart Personal Protective Equipment, Wearables and Internet of Things Technologies

Daniel Podgórski

CONTENTS

6.1 INDUSTRY 4.0 AND ICT-BASED MANAGEMENT OF OSH

The global economy is driven by global competition and the need to rapidly adapt production processes to quickly changing market needs and requirements. These requirements can only be achieved through radical advances in manufacturing technologies, often referred to as the digital transformation of industry, as they aim at a profound integration of digital technologies and business processes leading to new business models.

Meeting the challenges of the digital transformation lies at the heart of the so-called fourth industrial revolution, which is often labelled as Industry 4.0, and is characterised first of all by the widespread automation of industrial processes, including the extensive use of industrial and collaborative robots, additive manufacturing technologies, the concepts of industrial Internet of Things (IIoT), cyber-physical systems (CPS), cloud and edge computing, as well as the use of advanced data analytics methods, including machine learning and the dimension of big data.

The term Industry 4.0 was used for the first time in 2011, to describe a joint initiative of representatives of business, government and academia within the "High-Tech Strategy 2020 for Germany" concerning the promotion of computerisation of manufacturing and applications of digital technologies in industrial sector in order to strengthen the competitiveness of German industry (Kagermann et al. 2011). This term is now widely used worldwide and has been successfully adapted to shape and promote new industrial policies in many countries.

The central paradigm of Industry 4.0 is the introduction of the idea of the smart factory, a factory that will be equipped with the context-awareness functions and may assist people and machines in a holistic way in execution of their tasks. Such a factory is to provide the possibility of product customisation considering individual and changing customer needs, as well as making possible rapid adaptation to market changes and emergency situations, ensure efficiency of the use of resources and energy, building new models of co-operation with partners and extending the production process to suppliers and customers.

With the development of the above-mentioned concepts and digital production technologies, there have also been attempts to develop and apply ICT applications in the field of OSH, in order to ensure an adequate level of safety and health in workplaces functioning within the processes of smart manufacturing. This includes, in particular, the use of new sensor technologies, which offer opportunities to improve the safety and health of workers by real-time monitoring of various harmful and noxious parameters of the working environment. In addition, the use of new materials, sensors and telecommunications technologies in combination with cloud computing and big data analytics methods makes possible the introduction of other functions that are key to the effective management of OSH, such as the identification of hazards and reducing the associated risks in real time, detection of unsafe workers' behaviours, automatic incident detection and recording, monitoring of the effectiveness of individual OSH management processes, identification of workers' needs with regard to training and skills improvement in specific OSH fields, and many others.

Against that background, the purpose of this chapter is to provide an overview of the digital transformation issues to demonstrate the potential of the newest technologies to manage safety and health at work, which is especially important given the global trends in the development of intelligent manufacturing systems. In particular, this chapter will review the roles of smart personal protective equipment (smart PPE) and other smart wearable devices for workplace applications, with a particular emphasis on the usefulness thereof in the context of IT-empowered OSH risk management. These issues will be complemented by an analysis of the potential application of big data analytics for OSH management, as well as by discussing the inter-related aspects of privacy, personal data protection and cyber-security, which are becoming more and more important, and to some extent a hindrance to the more widespread dissemination and application of otherwise useful digital technologies in the workplace.

6.2   THE DEVELOPMENT OF SMART PPE TECHNOLOGIES AND WEARABLE DEVICES FOR USE IN THE WORKING ENVIRONMENT

6.2.1   DEFINITION AND ROLES OF PERSONAL PROTECTIVE EQUIPMENT IN PROTECTING WORKERS AGAINST HAZARDS AND RISKS

Personal protective equipment (PPE) is usually defined as "equipment designed and manufactured to be worn or held by a person for protection against one or more risks to that person's health or safety" (European Union 2016a). PPE items are often divided into groups and types on the basis of parts of the human body to be protected and with regards to the risks they provide protection against. The commonly used PPE typologies divides these products into the following main groups:

- protective clothing (including PPE signalling the user's presence visually, and lifejackets);
- protection against falls from height (full body harness, waist belts, fall arresting devices, anchor devices, etc.);
- hearing protection (earmuffs and earplugs);
- eye and face protection (goggles, safety glasses, face shields);
- respiratory protection (filtering devices, insulating devices, breathing equipment);
- hand and arm protection (safety gloves, elbow pads);
- foot and leg protection (footwear, instep protectors, kneepads);
- skin protection (barrier creams).

In accordance with the classical hierarchy of occupational risk controls PPE is considered as a last resort in risk reduction, because it should be taken into account only when higher level types of controls are not feasible or effective to limit the risk to an acceptable level. Despite the fact that this hierarchy does not arise from a specific research base, it has been widely accepted by OSH professionals and presented in a similar way in many national and international standards, guidelines and legal requirements concerning OSH management (Manuele 2005). For example, international standard ISO 45001:2018 (ISO 2018), which defines specifications for OSH management systems, refers to the hierarchy of risk controls (clause 8.1.2) in the following way:

The organisation shall establish, implement and maintain a process(es) for the elimination of hazards and reduction of OH&S risks using the following hierarchy of controls:

a) *eliminate the hazard;*
b) *substitute with less hazardous processes, operations, materials or equipment;*
c) *use engineering controls and reorganisation of work;*
d) *use administrative controls, including training;*
e) *use adequate personal protective equipment.*

Despite the convincing logic behind the above-mentioned hierarchy, as well as several decades of its application in practice, the use of PPE is still one of the most effective methods of reducing OSH risks in many industrial sectors and workplaces. This is particularly true for workplaces with complex and harsh environmental conditions exposing workers to a wide range of life and health risks at the same time, and for workplaces where risk factors change dynamically in an unforeseen way, thus hindering or preventing the use of rationally designed protective measures representing the higher levels of risk controls hierarchy.

6.2.2   Smart PPE-Enabling Technologies

The dynamic development of technologies in such domains as materials engineering, electronics, computer science, and wireless telecommunication, which have taken place over the last few decades is characterised by the striving for miniaturisation of devices, the increasing mobility of computing systems, as well as the development of distributed systems constituting a set of independent smart objects and technical devices combined into one logically coherent whole (the Internet of Things [IoT]). At the same time, there is better and better mastering of sensor technologies that enable one to sense (i.e. to detect) some physical, chemical or spatial characteristics of real-world objects and environments. As a result, a number of advanced concepts and enabling technologies are available, which offer practically innumerable possibilities of creating and improving various smart PPE solutions.

Enabling technologies belonging to the field of functional and smart textile materials deserve particular attention, as this field covers a wide range of different material technologies that make possible specific functions obtained by means of material composition, construction and/or finishing, and which are based on the use of various physical and chemical phenomena. Taking these phenomena as the basis for the typology, as proposed in the CEN/TR 16298 technical report (CEN 2011), the types of functional textile materials include textile materials that are electrically conductive, thermally conductive, thermally radiative (emissive), optically conductive, fluorescent, and phosphorescent, as well as textile materials which release specific substances. Smart textile material can be defined as a functional textile material which interacts actively with its environment, i.e. responds or adapts to changes in the environment.

In addition to material technologies, the development of smart PPE has also been driven by significant achievements in other contemporary engineering disciplines, such as:

- health monitoring micro- and nano-sensors integrated with textiles;
- flexible, stretchable and conformal electronics (elastronics);
- MEMS (micro-electro-mechanical systems) and LOCs (lab-on-a-chip);
- ultra-low power electronics and energy harvesting technologies (thermo-electric, piezoelectric, photovoltaic, electrodynamic, etc.);
- wireless telecommunication technologies and protocols, such as RFID, iBeacons, Wi-Fi, NFC, Bluetooth, BLE (Bluctooth Low Energy), ZigBee, Z-Wave, UWB, etc.;
- augmented- and mixed-reality technologies;
- cloud computing* and edge computing† applications; and
- machine learning algorithms and big data analytics.

---

* Cloud computing is the on-demand availability of computer system resources, especially data storage and computing power, without direct active management by the user. The term is generally used to describe data centres available to many users over the Internet (Wikipedia: https://en.wikipedia.org/wiki/Cloud_computing, accessed August 6, 2019).

† Edge computing is a distributed computing paradigm which brings computation and data storage closer to the location where it is needed, to improve response times and save bandwidth (Wikipedia: https://en.wikipedia.org/wiki/Distributed_computing, accessed August 6, 2019).

In particular, the last two of the areas mentioned above deserve greater attention, as the development of applications based on cloud and edge computing technology has made possible a significant increase in the amount of data that can be collected and processed in smart PPE systems in real time. Whereas, the application of machine learning algorithms, including the use of methodologies making possible the processing of massive amounts of data (i.e. big data), enable one to obtain actionable insights, which can very useful for the introduction of advanced functions in the area of OSH. These issues will be explained and illustrated in more detail in the other sections of this chapter.

The general fundamental feature of smart textile material, which is the intrinsic ability for automatic reaction to changes in the surrounding environment, can be used to define the entire category of smart PPE systems that demonstrate the same feature. Taking this approach into account the following definition was proposed:

> A smart personal protective system (SPPS) is an assembly of devices or elements that is designed to be worn or held by a person for protection against one or more health and safety hazards, and that reacts automatically, either to changes in its environment or to an external signal. (Marchal and Baudoin 2018)

6.2.3 Workplace Wearable Devices

When considering the use of digital technologies for managing OSH, account should also be taken of the role of other electronic devices that can be worn by workers, commonly known as "wearables" or "wearable devices". These devices do not provide any direct protective functions in the way that traditional PPE do, but their measurement, monitoring and control functions can be successfully applied to detect hazards and reduce risks in the workplace (Choi et al. 2017; Awolusi et al. 2018; Dolez et al. 2018; Barata and da Cunha 2019).

According to the definition proposed in draft standard IEC 63203-101 ED1 (IEC 2019a) wearable electronic devices constitute an "electronic device intended to be located near to, or on a human body. These devices can be provided with artificial intelligence functionality and/or can be part of a widely connected system. These devices can be used to perform a variety of tasks". Investopedia (2020a) defines these devices in more concrete terms as a:

> category of electronic devices that can be worn as accessories, embedded in clothing, implanted in the user's body, or even tattooed on the skin. The devices are hands-free gadgets with practical uses, powered by microprocessors and enhanced with the ability to send and receive data via the Internet.

Wearable devices used in the workplace can be classified taking into account their physical form, location and manner of wear, as well as functions based on various physiological parameters, working environment parameters and other data generated by machines or processes operated by workers. For example, one of the classifications proposed by Mardonova and Choi (2018) distinguishes the following types of wearables: smartwatches, smart eyewear (smart glasses), fitness trackers, smart clothing, wearable cameras and wearable medical devices. Wearables can also be classified according to the location where they are worn or fixed to the body of a person or the clothing. In this case, such location areas as the head, ears, eyes, torso, arms, wrists, legs, and feet can be distinguished. But wearables may also be implantable into various parts of human body, and be multifunctional, i.e. they can be worn on different parts of the body depending on the user's needs.

The main areas of wearables applications with benefits for employers and workers include, among others, monitoring psychological and physiological factors, enhancing operational efficiency, collaborating, promoting work environment safety and security, performing industrial designing, and improving the health of workers (Khakurel et al. 2018). Among many cases of practical application of non-PPE wearables to OSH-related purposes that have been described in the literature, the following two examples can be mentioned: applying the Philips Health watch for monitoring the physical activity of workers within the We@Work project (Abtahi et al. 2017), and a study on the implementation of the Microsoft HoloLens® mixed-reality device at construction sites to improve work-related hazard identification and risk communication (Dai and Olorunfemi 2018).

6.2.4   SMART **PPE** FUNCTIONS AND TYPOLOGY

As mentioned above, recent developments in the fields of enabling technologies has allowed PPE to be equipped with a number of new useful functions that did not previously exist in traditional PPE solutions. In particular, these functions include: monitoring the health of workers by measuring key physiological parameters (e.g. body temperature, heart rate, respiratory rate, etc.); monitoring work comfort at work (e.g. underclothing temperature and humidity, work position); geographical location of workers with regard to potentially dangerous physical objects or high-risk zones; monitoring the current status of PPE protective properties and the degree of wear; notifying workers and/or their safety managers about current hazards or the level of risk (e.g. by means of a personal digital assistant or smart glasses); and activation of protective systems or deactivation of sources of risks after exceeding a high-risk threshold value. The above-mentioned smart PPE functions, which are performed in addition to the basic protective functions, can be divided into three main types: (1) sensing, (2) activating risk controls, and (3) ensuring correct operation of PPE items, as shown in Table 6.1.

Depending on the number and nature of functions performed, the type of technology used to build smart PPE, and the extent to which these systems affect OSH in terms of the number of controlled environmental factors or the number of workers to be monitored and protected, three levels of complexity of these measures can be distinguished:

1) simple smart PPE;
2) autonomous smart PPE systems; and
3) smart networked PPE systems.

Such a division reflects both the diversity of smart PPE solutions from the point of view of their structural complexity and the level of intelligence that should be

TABLE 6.1

Main Types of Additional Functions Provided by Smart PPE

Sensing	Activating risk controls	Ensuring correct operation
• Monitoring hazards and risks in the working environment • Monitoring user's physiological parameters (state of health) • User location (e.g. proximity to machines, danger zones, etc.)	• Self-adjusting of protective properties • Activation of external risk control measures (engineering controls) • Providing warnings and/or work instructions to users (administrative controls)	• End-of-service-life indication • Damage detection and self-repair • Monitoring correct PPE application and usage • Energy harvesting and storage (to provide energy to power sensors, actuators and electronics)

adequate to the functions performed. The next sections provide a brief description and examples of solutions relating to each of these levels.

6.2.5 SIMPLE SMART PPE (LEVEL 1)

Level 1 covers relatively simple PPE solutions based on the use of smart textiles and other smart materials, such as phase-change materials (PCM), shape-memory alloys, thermoelectric materials and others. The basic functions of these solutions are to detect changes in the physiological parameters of users and the parameters of the surrounding environment and to react autonomously to these changes by adequately adapting the protective functions according to the general concept shown in Figure 6.1. Level 1 PPE products may also consist of some relatively simple analogue and digital electronic modules to provide dedicated functionality, but the processing of signals in such modules is usually limited to measuring a few input parameters, processing them according to the appropriate transformation functions, and then generating the output signals to achieve the desired state of protection against specific hazards.

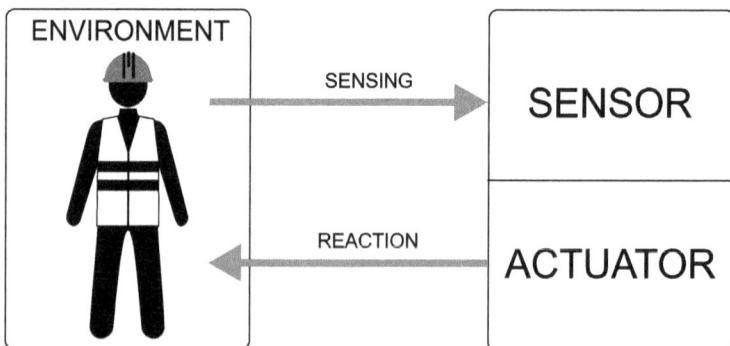

FIGURE 6.1   General concept of self-adaptive functions of simple smart PPE (level 1)

Level 1 smart PPE enables one to control only a single or, at the most, a few risks at a time, and their range of impact affects only individual users. Often these solutions are not ready-to-use products but are offered as to be considered as components for PPE items which provide traditional protection in a traditional manner, i.e. based on materials and devices that are not self-adaptive to environmental changes. Since these solutions do not contain ICT modules, they do not generate or transfer any data that could be used to perform more advanced functions within the overall OSH management system.

An analysis of the scientific literature and a review of commercial offers from PPE manufacturers show that there are already many different smart materials and smart PPE products on the market, which are able to effectively adapt protective parameters to changes in the working environment without the support of sensor technologies and the ICT. One of many examples of such solutions is self-thermo-regulating garments constructed of a material consisting of microcapsules containing phase-change material, which is designed to cool the body of a person working in a hot working environment or wearing impermeable protective clothing (Bartkowiak et al. 2013). Another example is a prototype of industrial safety helmet equipped with a cooling PCM layer and a simple electric fan-based system of forced convection to control the thermal comfort of users (Ghani et al. 2017).

6.2.6 Autonomous Smart PPE Systems (Level 2)

Level 2 corresponds to autonomous smart PPE systems, which are the intermediate solutions between level 1 and level 3 systems. Level 2 systems can consist of multiple embedded sensors, activators and control electronic modules. These systems can detect different hazards in the working environment and monitor risks associated with various physical, chemical, biological, etc. factors, but they can also monitor the proper functioning and end-of-service-life of individual system modules. At the activator level, the operation of level 2 systems involves primarily the adaptation of protective functions to mitigate the risks associated with detected hazards, but may also involve communicating the level of risk and related warnings to the worker, as well as triggering the operation of external risk reduction devices or systems that perform the functions of so-called engineering controls or, in equivalent terms, collective protection measures.(Figure 6.2).

The autonomy of the level 2 smart PPE means that they can properly perform all their essential protection functions without any connection to an external controller or data server (such as cloud-based systems). Such systems may be equipped with ICT modules for processing measurement data, but the collection, transmission and processing of these data will in this case be limited to the cyber-physical sphere of a single user, referred to as the Body Area Network (BAN), or possibly to the sphere of several smart PPE users operating locally near each other.

An example of level 2 smart PPE is a high-visibility smart vest equipped with sensors to detect workers approaching danger zones, and/or detect vehicles such as fork-lift trucks or automated guided vehicles approaching workers (Elokon 2019). The detection of potential collisions by such smart vest can trigger acoustic and visual alarm signals and send a signal to an approaching machine in order to limit its

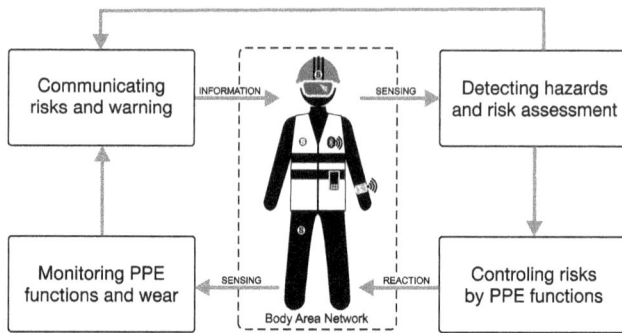

FIGURE 6.2 The concept and main functions of autonomous smart PPE systems (level 2)

speed or stop. Another example of a level 2 solution is the Dainese protective jacket which is equipped with fall detectors, inflating airbags and a pneumatic system activated automatically during a fall (Ridden 2017). The airbags are suitably positioned around the human body to provide protection for those parts of the body that are at risk of injury when falling from a height not exceeding two metres.

Despite discussing the above examples of smart PPE systems, level 2 systems should be considered as a certain thought construct or transitional stage, which is introduced in order to organise and describe the complexity and role of various smart PPE solutions in OSH management, rather than a clearly identified category of smart PPE represented by a set of products introduced for use in specific workplaces. In view of the rapid development of industrial IoT technology, in particular wireless communication technology and edge- or cloud-based data processing services, and the increasing use of non-PPE wearables in the workplace, level 2 applications are not found frequently in the literature, as they are actually a special case of the more complex and effective level 3 systems that are described in the next section.

6.2.7   SMART NETWORKED PPE SYSTEMS (LEVEL 3)

Smart networked PPE systems are a specific type and area of application of smart networked systems (SNS) dedicated to different working environments to improve the safety, health and/or comfort of workers. SNS can generally be defined as collection of spatially and functionally distributed embedded computing nodes that are interconnected by means of wired or wireless communication infrastructures and protocols (Podgórski et al. 2017). Such systems consist of a large number of interconnected sensors, activators and ICT modules, can be connected with external (e.g. cloud-based) or local processing servers, and on this basis can provide real-time collection, processing and analysis of large measurement data streams from multiple users and other smart objects covered by an industrial IoT network. The concept of such a system is shown in Figure 6.3.

As shown in Figure 6.3, level 3 systems can, to some extent, be considered as a network of interconnected level 2 systems. Such infrastructure can also include a master node that conducts coordinative functions and calculations for the whole system. Thus, level 3 smart PPE systems can simultaneously control many different

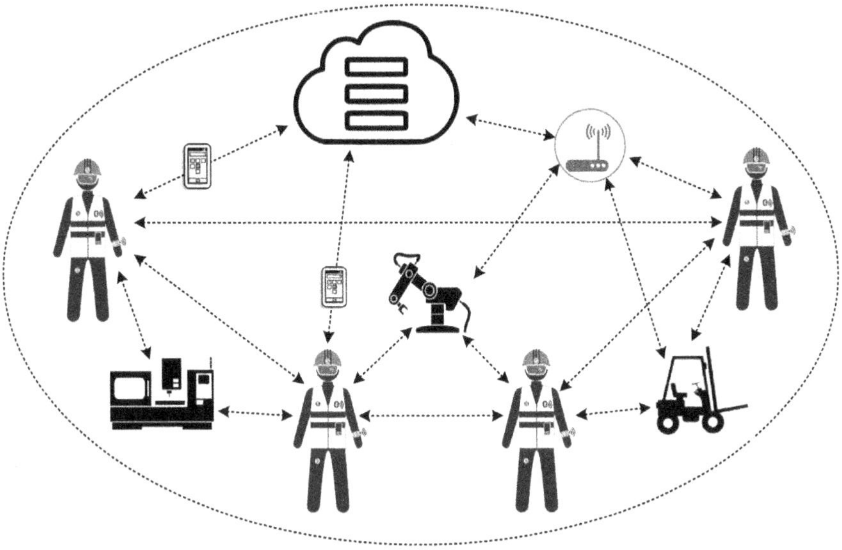

FIGURE 6.3 General concept of smart networked PPE systems (level 3)

risk factors in the working environment, risk control functions may cover many users at the same time, and, in the case of using advanced data analytics and context-awareness methods, these systems can support the implementation of advanced OSH management functions. The extended range of functions of smart networked PPE systems makes possible their integration into industrial IoT covering production machines and devices, industrial robots, collective protective systems and other smart objects existing in a given working environment.

Such advanced smart PPE systems representing the third level of system complexity are not yet widely used in the workplace, but more and more research centres and high-tech companies are carrying out research and pilot testing in this area. As a result, the first commercial offerings of such systems are already available on the market. One example is the Corvex Connected™ platform developed and offered by Corvex Connected Safety (Webb 2019). This technology enables data to be transferred between the smart PPE items worn by workers and a network of beacons placed at different locations in the workplace. The area monitored by this system can be divided into appropriate zones where different requirements with regard to safety and health and the use of PPE may apply. Each worker is equipped with a unique Corvex handheld device that allows real-time notification of current risks at their workstations, as well as providing instructions on how to use the appropriate PPE.

Another example of the implementation of the smart networked PPE system is Maximo Worker Insights application developed by IBM on the basis of the IBM Watson platform in co-operation with Garmin Health, Guardhat, Mitsufuji and SmartCone (Vinoski 2019). The aim of this platform is to monitor such potential hazards occurring in the work environment as heat, temperature, gas levels, and working at height, and to assess the associated risk on the basis of measurement data collected from the workplace and the sensors worn by workers. Biomonitoring of workers is going to be provided by Garmin's fitness tracker and Mitsufuji's smart shirt. The Guardhat safety helmet, on the other hand, will make it possible to monitor the physical conditions around the worker and warn him/her about the detected hazards. Finally, SmartCone technologies are going to enable the separation of safety zones in the workplace and the measurement of environmental noise and temperature levels. It is claimed that the potential advantage of this system is the ability to use the high analytical power offered by the IBM Watson platform for analysis and multifaceted inference based on large amounts of data obtained simultaneously from many different sources of the working environment with the use of various measuring devices and technologies.

6.3   ICT-EMPOWERED OSH RISK MANAGEMENT PROCESS

6.3.1   New Properties of the OSH Risk Management Process in the Context of ICT-Based Smart PPE System Implementation in the Workplace

Performing occupational risk assessments in the workplace, and then introducing appropriate protective and preventive measures to reduce the risk to an acceptable level, is widely recognised as a key process for an effective OSH management system in the enterprise (ILO 2001; OSHwiki contributors 2019), since the efficiency of this process has a direct impact on preventing accidents at work, occupational diseases and related economic losses for both workers and society. As shown in the previous section, the new functions introduced by smart PPE systems and other workplace wearables make possible faster and more effective detection (in practice, in real or near-real time) and then, as a consequence, a reduction of OSH risk factors in the working environment to an acceptable level. This applies in particular to functions such as measuring and monitoring workers' exposure to harmful environmental factors (physical, chemical, etc.), monitoring workers' health by measuring key physiological parameters (e.g. body temperature, heart rate, respiratory rate, etc.), monitoring work comfort (e.g. underclothing temperature and humidity, work position, etc.) and geographical location of workers with regard to potentially dangerous physical objects or high-risk zones, monitoring the current state of PPE protective properties, communicating to workers and/or safety managers about detected hazards and risks, self-adjusting of protective properties of given PPE items, and/or activation of external risk control measures. These functions allow to apply a better approach to OSH risk assessment and management processes, which so far have often been performed in a rather formal manner, which is mainly due to the obligation to meet legal requirements concerning safety and health at work.

In static workplaces, i.e. those where working conditions do not change dynamically, the risk assessment for individual workplaces is carried out in practice no more frequently than every few months, which means that the risk assessment results obtained at a given moment are considered to be averaged over a longer period of time. In addition, it is a common practice to carry out risk assessments for groups of workers, e.g. by grouping together workers exposed to the same or a similar set of harmful agents or by grouping together similar workstations where tasks are performed in the same location and under the same conditions. Therefore, information on the results of risk assessment documented in the enterprise usually indicates a level of risk, which is a generalised value for previously defined groups of workers or for sets of identical or similar jobs.

In the context of this, the use of smart networked PPE systems for OSH risk management make possible the use of this process in real or near-real time, and the personalisation of the outcomes of these process in relation to individual workers. The first of these features is of great importance due to the increasing dynamism of changes in the work environment, resulting from, inter alia, the introduction of new business models and new production concepts, e.g. within the framework of smart factories. On the other hand, the personalisation of OSH risk management makes it possible to focus on the safety and health conditions of individual workers who, being exposed at a given moment to various factors of the working environment, can be protected against the impact thereof in a differentiated manner that is adapted to their needs depending on the type of hazard and the current level of exposure.

6.3.2 GENERIC MODEL OF THE ICT-EMPOWERED OSH RISK MANAGEMENT PROCESS

Taking into account the functions of smart networked PPE systems as well as the concept of integration of these systems with other networks and systems of the industrial IoT, especially in the context of the development of Industry 4.0 concepts, the author proposed a graphic representation of a general model of a cyber-physical system supporting OSH risk management process in smart working environments, which was originally published in Podgórski et al. (2017). Figure 6.4 shows a revised version of this model, which demonstrates the basic functionalities of the system and takes into account new aspects resulting from the development of smart PPE systems, workplace wearables and the industrial IoT.

The hardware layer of the presented model consists of sensors and actuator networks embedded in workplace equipment and smart PPE and wearables worn by workers, as well as other sensors and actuators embedded in machinery and other workplace objects and facilities. Whereas the software layer can be divided into four main parts: (1) a contextual database containing historical and current measurement data collected from all sensors integrated or associated with different physical objects; (2) data fusion and context-aware aggregation module; (3) an inference engine responsible for analysing contextual data and calculating and allocating risk levels to individual workers in real time; and (4) a risk control manager that is responsible for analysing the resources available to control risk, selecting the appropriate preventive and protective measures in view of their functions and potential effectiveness, activating those measures in due time, and monitoring their use.

The proposed model points to the crucial role of context-awareness in ICT-supported processes of hazard detection and risk assessment. The concept of context used for ICT applications can be defined as "any information that can be used to characterise the situation of an entity. An entity is a person, place or object that is considered relevant to the interaction between a user and an application, including the user and applications themselves" (Dey 2001). Whereas context-awareness

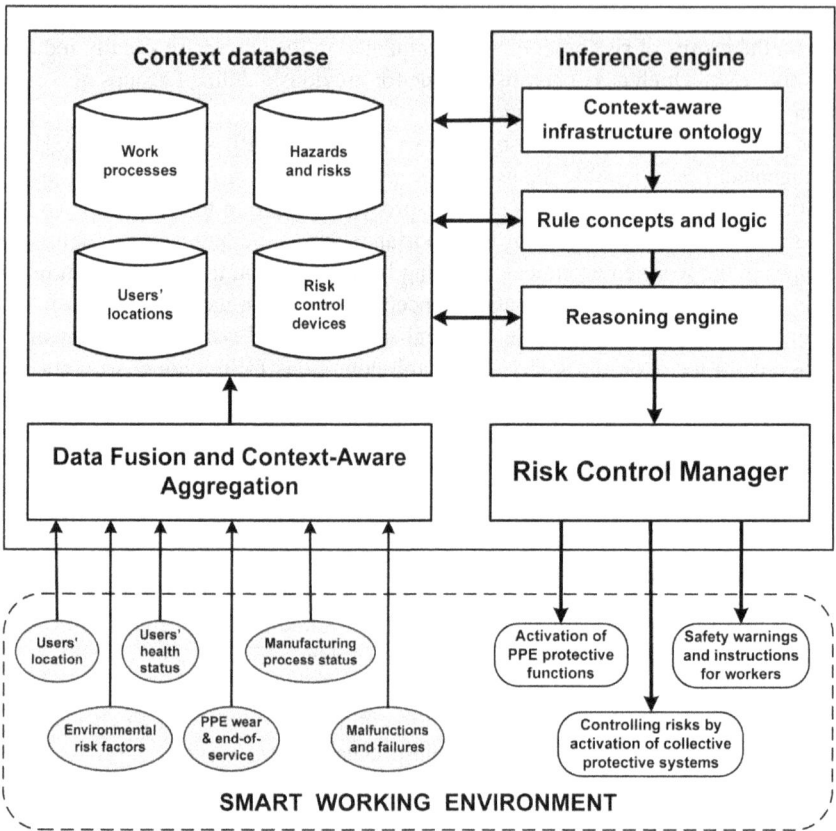

FIGURE 6.4 Generic model of a cyber-physical system applied for OSH risk management in smart working environments (modified from Podgórski D. et al. [2017]. Towards a conceptual framework of OSH risk management in smart working environments based on smart PPE, ambient intelligence and the Internet of Things technologies. *International Journal of Occupational Safety and Ergonomics*, 23(1):1–20)

and context-based reasoning are basic features and functionalities of cyber-physical systems implemented in smart environments facilitated by IoT technologies (Hong et al. 2009; Bettini et al. 2010; Ye et al. 2012). This applies in particular to environments characterised by frequent and dynamic changes that significantly affect the behaviour of users and their interactions with other objects in those environments.

6.3.3   REVIEW OF PROBABILISTIC METHODS FOR CONTEXT REASONING AND OSH RISK ASSESSMENT

As it has already been mentioned in section 6.3.1, the most significant functionality of the ICT-based smart networked PPE systems is the capability of real-time and personalised assessment of occupational risks, taking into account all hazardous factors of a dynamically changing production environment. There are many methods and techniques of work-related risk assessment that have been developed so far – see for example reviews by Marhavilas et al. (2011) and Wang et al. (2015). In general, risk assessment techniques can be divided into two types consisting respectively in the deterministic and the probabilistic (stochastic) approaches (Kirchsteiger 1999; Marhavilas and Koulouriotis 2012). Next, deterministic techniques can by grouped in three main categories: (a) qualitative, (b) quantitative, and (c) hybrid techniques (qualitative-quantitative, semi-quantitative), while stochastic methods are roughly divided into the methods based on two categories: (a) the classic statistical approach, and (b) accident forecasting modelling (Marhavilas and Koulouriotis 2012). It is also claimed that the application of a single risk assessment method will not ensure sufficiently reliable results, so it is advisable to use mixed approaches based on both deterministic and probabilistic methods. At the same time an opinion should be taken into account that probabilistic methods may have an advantage over the deterministic ones, because the former are more cost-effective and their results are easier to communicate to users and other stakeholders (Kirchsteiger 1999).

The literature concerning ICT applications in the field of smart environments and pervasive computing reveal that in the case of relatively simple systems, which consist of up to several sensors and actuators (e.g. location tracking and warning system for workers operating in dangerous zones), the methods for risk assessment and reasoning are usually rule-based and have a deterministic nature. One may find a few examples of such applications in papers by Giretti et al. (2009), Teiser et al. (2010), Yang et al. (2011), Ku and Park (2013), and Fugini and Teimourika (2014). However, in the case of more advanced cyber-physical systems that consist of many sensors monitoring various parameters of users' health and the environment, and which are used for managing complex scenarios of users' activities and their relations with smart objects, the use of probabilistic methods or the use of methods based on fuzzy logic is needed (Ranganathan et al. 2004; Haghighi et al. 2008; Singla et al. 2009). In particular this approach is necessary in scenarios where a certain level of uncertainty occurs, which is related to, for example, the recognition and prediction of user behaviour, inaccuracy of sensors, missing information, imperfect observations and inferring on the basis of imprecise and conflicting data. In such cases literature often indicates the practical usefulness of such methods as Baeysian networks, Hidden Markov Models and Dempster-Shafer theory of evidence (Tolstikov et al. 2011; Ye et al. 2012; Perera et al. 2013).

Baycsian networks (BNs) are used to model cause-and-effect relationships between random variables and their conditional dependencies, thus providing a concise representation of their joint probability distribution. Modelling the dependency is carried out by constructing a directed acyclic graph, where nodes represent random variables and the edges correspond to relationships between these variables. BNs have found many practical applications in many domains of science, economy and life, which is reflected in some reviews, e.g. by Kenett (2012) and Weber et al. (2012). Since the BNs are a good tool to calculate the probability of occurrence of various inter-related random events, they have become a frequently used tool for risk estimation, including the risk associated with work (e.g. Leu and Chang 2013). They can also be used for other cause-and-effect analysis in the framework of OSH management, such as analysis of accidents caused by falls from a height

(Martín et al. 2009), or in investigations on relationship between hygiene conditions, ergonomic conditions, job demands, physical symptoms, psychological symptoms, and occupational accidents (García-Herrero et al. 2012). Examples of existing BN application for context reasoning in smart environments include, among others, environmental situation recognition in wireless sensor networks applied for outdoor environment monitoring (Bagula et al. 2010), context reasoning for fall risk assessment of elderly in ambient assisted living (Koshmak et al. 2014), and a real-time monitoring and early warning system that safeguards the well-being of workers exposed to heat stress (Liu and Wang 2017).

Hidden Markov Models (HMM) are a statistical tool in which it is assumed that the modelled system is represented as a Markov chain with hidden states, but with the visible outputs, which are dependent on these states. In turn, a Markov chain is a model of a stochastic process constituting a series of events in which the probability of each event depends only on the outcome of the preceding event. HMM, similarly to BNs, have wide applications in many various domains of science and life, as for example in biology (Choo et al. 2004), speech recognition (Gales and Young 2007), computer vision and pattern recognition (Fink 2014), and many others. In the area of context inferring in smart environments HMM have been applied among others for modelling human behaviour in smart hospitals (Sánchez et al. 2008), and smart homes (Bruckner and Velik 2010; Chahuara et al. 2013). Examples of practical HMM applications in the working environment include location tracking and activity analysis of construction workers (Khosrowpour et al. 2014), predicting the likelihood of work-related musculoskeletal hazards among dental students (Thanathornwong et al. 2014), and real-time location tracking, trajectory prediction, and prevention of potential collisions between workers and construction site objects (Rashid et al. 2018).

The third of the above-mentioned probabilistic method is the Dempster-Shafer theory of evidence (DST), also referred to as the theory of belief functions. DST was initially developed by Dempster (1966), and was then extended, refined and recast by Shafer (1976). This method is regarded as a generalisation of the Bayesian theory of subjective probability, since in traditional probability theory evidence is associated with only one possible event, while in DST, evidence can be associated with multiple possible events. In general, DST consists of the calculation of what is known as the "belief function" in order to obtain degrees of belief by combining all available information (evidence) from different sources. DST has been used to solve problems of processing incomplete or uncertain information in business decision making (Srivastava and Liu 2003) and in many engineering disciplines, such as computer science, construction and production management. In the field of smart environments DST is often used to support data fusion and/or for context reasoning and human behaviour recognition in smart homes (e.g. Zhang et al. 2010; Sebbak et al. 2013), while examples of applying DST for risk assessment and safety management have been presented by, for example, Rakowsky (2007), Zhang et al. (2014), and Nesculescu et al. (2015).

6.3.4 NEW ROLES OF SMART PPE IN THE HIERARCHY OF OSH RISK CONTROLS

The use of new functions of smart PPE and wearables in combination with ICT applications to support real-time OSH risk management throws a slightly different light on the role of these devices in relation to the classical hierarchy of risk controls, which has been outlined in section 6.2.1. In literature, this hierarchy is often presented graphically, in the form of an inverted pyramid illustrating the relative effectiveness of individual measures for accident prevention and the protection workers' health. This approach comes from several decades ago and thus reflects a static and periodic approach to risk management resulting from the technological capabilities and organisational practices available in this field at that time. Since the situation in this area in the last decade has changed significantly, mainly due to the development and implementation of the Industry 4.0 concepts and dynamic progress in the development of smart PPE, sensor technologies and wireless ICT networks, it is appropriate to consider the new interpretation of this hierarchy by taking into account the new functions offered by smart PPE systems.

According to the discussed hierarchy elimination and substitution are considered to be the most effective and at the same time the most difficult risk control measures to be applied in existing work processes. These measures are justified mainly at the stage of designing and developing a technology process, a workplace or a service, during which it is still possible to introduce relatively easy and low-cost changes consisting of the elimination of hazard sources or replacement of materials generating hazards with low-emission materials or materials of chemical composition safe for humans. However, in the case of already introduced technological processes, the introduction of risk controls such as elimination or substitution may be unrealistic or economically unjustified, as it may mean the necessity to stop the production process for a longer period of time in order to completely redesign it. Similar difficulties are also to be considered in the case of designing work processes, which are introduced within the framework of digitally transformed smart factories. Therefore, these two highest levels of risk controls, namely elimination and substitution, can also be considered adequate in the case of ICT-based OSH risk management, with the reservation, however, that these activities should be taken into account primarily at the stage of designing new processes.

In traditional non-smart workplaces, the interpretation of the priorities for risk control measures belonging to lower levels of the hierarchy is quite clear. Engineering controls (referred to also as collective protective equipment) should be used first to separate workers from exposure to hazards. If, despite these actions, it is not possible to reduce risk to an acceptable level, administrative controls should be applied (e.g. changes in work procedures, limiting working time at selected positions, additional training or instructing workers, providing warnings to workers on existing hazards, etc.). And only if these earlier actions did not bring the desired effects, should workers be provided with PPE (as so-called the "last resort"), provided that the respective PPE items are appropriately selected and matched to the types of hazards and risks occurring at their workplace.

Such hierarchy of protective measures should be respected in traditional workplaces at the stage of designing workstations and processes, and at the operational stage at the time of planning risk control measures, e.g. those resulting from periodic or ad-hoc risk assessments. However, in workplaces where there are frequent and significant changes in the exposure of workers to various types of risks, strict compliance with these principles may lead to selection of less effective risk controls,

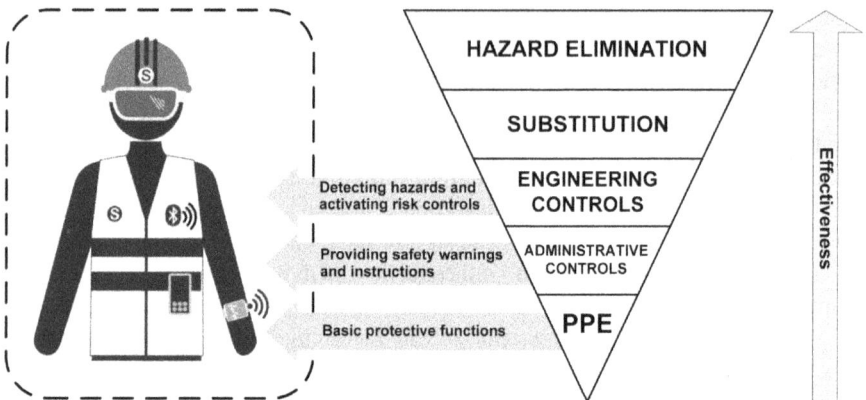

FIGURE 6.5 New functions of smart PPE systems in relation to the classical hierarchy of OSH risk controls (adapted from Podgórski D. [2017]. Functions of smart PPE in relation to the Hierarchy of Risk Controls, LinkedIn. Posted September 11, 2017)

where other, readily available functions offered by smart PPE systems would provide a better level of protection. In such cases, it seems appropriate to consider other priorities for protective measures in line with the interpretation of the risk control hierarchy presented in Figure 6.5.

The proposed interpretation of the hierarchy of risk controls means that if an employer has an access to smart PPE or other wearable devices that have novel functions allowing to control given risks more effectively, e.g. by initiating appropriate engineering controls or performing various types of administrative controls, the use of these smart devices may have a higher priority comparing to the case of traditional non-smart PPE, which should always be considered as the "last resort".

6.4   PERSPECTIVES FOR APPLYING MACHINE LEARNING AND BIG DATA ANALYTICS TO OSH MANAGEMENT PROCESSES

As mentioned in section 6.2.2, the progress observed recently in the field of ICT, and in particular the development of applications based on cloud and edge computing, makes it possible to collect and process more and more data, including that collected through smart PPE and workplace wearables, which make it possible to monitor in real time the state of working conditions and behaviour of workers in the context of safety and health at work. On the other hand, the growing number of data sets obtained by these systems makes it possible to use artificial intelligence methods, including machine learning and big data analytics algorithms, to obtain new knowledge and useful insights suitable for advanced OSH management functions. The purpose of this section is to present some leading concepts in this field, review selected examples of their application in practice, and discuss perspectives for their further development and use to support key processes in OSH management systems.

6.4.1 BASIC NOTIONS AND UNDERLYING CONCEPTS

From the very beginning of the development and promotion of the concept of digital transformation of industry, both in the professional literature dealing with new industrial technologies and in the scientific literature on computer science and related domains, as well as in other media, the terms such as "artificial intelligence", "machine learning", "deep learning", "big data analytics", "predictive analytics", etc. began to appear more and more frequently. These concepts are, to a great extent, inter-related and often used interchangeably, therefore, in order to properly discuss them, their brief definitions are presented below.

Artificial intelligence (AI) is defined in literature in many ways, depending on the needs of the authors or users of a given definition. For example, the Merriam-Webster Dictionary (2020a) defines this as: "1: a branch of computer science dealing with the simulation of intelligent behaviour in computers, 2: the capability of a machine to imitate intelligent human behaviour". The Council of the Organisation for Economic Co-operation and Development (OECD) gives the following definition: "An AI system is a machine-based system that can, for a given set of human-defined objectives, make predictions, recommendations, or decisions influencing real or virtual environments. AI systems are designed to operate with varying levels of autonomy" (OECD 2019). The second definition refers to the development and deployment of "intelligent" systems and therefore seems to be more appropriate to have in mind when reading this chapter.

Machine learning (ML) is a dynamically developing sub-area of AI, which deals with algorithms for analysing and learning from data and then using the acquired knowledge to produce useful insights and/or predict the course of various phenomena in the future. Machine learning algorithms are commonly divided into three main categories, i.e. (1) supervised learning, (2) unsupervised learning and (3) reinforcement learning, although many scholars propose more complex taxonomies. For example, Ayodele (2010) extends this division into three additional categories, semi-supervised learning, transduction, and learning to learn, while Dey (2016) also lists another four: multi-task learning, ensemble learning, neural networks, and instance based learning. Describing and discussing the different types and possibilities of machine learning algorithms is outside of the purpose and scope of this chapter, but there is a wealth of literature and many Internet resources where interested readers can find more knowledge in this field, e.g. in Ayodele (2010), Dey (2016), Kelleher et al. (2015), and Kubat (2017).

Deep learning is in turn a sub-field of machine learning that can be defined as "an artificial intelligence function that imitates the workings of the human brain in processing data and creating patterns for use in decision making. Deep learning is a subset of machine learning in artificial intelligence (AI) that has networks capable of learning unsupervised from data that is unstructured or unlabelled. Also known as deep neural learning or deep neural network" (Investopedia 2020b). Deep learning methods make it possible to solve complex cognitive problems such as object detection, speech recognition, translation of texts into other languages, recognition and classification of images, etc.

Predictive analytics is a form of advanced data analytics that aims to extract data from analysed data sets in order to identify specific trends and patterns that can then be used to predict future trends or events. The techniques of predictive analytics can generally be divided into two main groups: the first is aimed at discovering histori-cal patterns in the outcome variables and extrapolate them to the future, while the second is aimed at capturing the interdependencies between outcome variables and explanatory variables, and exploit them to make predictions (Gandomi and Heider 2015). Predictive analytics may be applied to both small and very large datasets.

Big data refers to high-volume, high-velocity and/or high-variety information assets that demand cost-effective, innovative forms of information processing that enable enhanced insight, decision making, and process automation (Gartner IT Glossary 2020a); however, there are many similar, although differently formulated, definitions of this term in literature. Big data was initially described by indicating three main determinants of this concept, often referred to as three Vs (Gandomi and Heider 2015), i.e. *volume* (amount of data, e.g. in terms of terabytes or petabytes), *variety* (i.e. structured and/or unstructured, variety of data forms), and *velocity* (speed of data generation, real time, near time or off-line). But recently, some authors and influencers from the field of big data analytics and related domains (e.g. Marr 2014) mention two other Vs as important features of big data, namely *Veracity*, which refers to the messiness or trustworthiness of the data, and *Value*, which emphasises the need to collect such data, which will have business value, and the benefits of their processing will be greater than the costs incurred.

Big data analytics concerns the application of specific methods for the analysis of data sets corresponding to the above-mentioned definition of big data, for which the analysis with the use of traditional statistical methods is pointless or impossible. The aim of big data analytics is to obtain useful actionable insights that can be used for making decisions leading to the development of new products and services, improving the efficiency of processes, reducing the costs of business and generally to acquire new knowledge, which can then be used for many different purposes. The process of big data analytics can be roughly divided into two stages: data manage-ment and data analytics (Gandomi and Haider 2015). The data management includes data acquisition and recording, data extraction, cleaning, and annotation, as well as data integration, aggregation and representation. The second stage consists of data modelling and analysis and is followed by the interpretation of data.

In the literature and practice on AI, machine learning and big data analytics one can often find another term, "data mining", which can be defined as "the pro-cess of discovering meaningful correlations, patterns and trends by sifting through large amounts of data stored in repositories. Data mining employs pattern recogni-tion technologies, as well as statistical and mathematical techniques" (Gartner IT Glossary 2020b). This definition points to close inter-relations between this concept and the ones presented above, as, for example, machine learning and deep learning techniques can be used in favour of data mining, while deep mining methods can be a part of prediction and big data analytics methodology.

6.4.2 Applications of Machine Learning and Big Data Analytics in the Field of OSH

Dynamic development of AI methods and technologies, including, in particular, increasingly better techniques and tools in the field of machine learning and big data analytics, making possible the acquisition of new knowledge and useful insights, both on the basis of historical data and data acquired on an ongoing basis, made it possible to apply these technologies in practice in many different sectors of business activity, including the support of activities in the area of OSH (Ajayi et al. 2018). This section presents some examples of applications based on machine learning and big data analytics in off-line mode, i.e. those that aim at acquiring new knowledge which can later be used by employers, safety managers and other stakeholders to better plan and perform OSH-related activities. The next section will be devoted to the application of artificial intelligence combined with ICT and IoT networks to support selected OSH management processes to be carried in real or near-real time.

One of the first achievements in this field is the use of supervised learning algorithms (SVMs) to predict accidents at work on the basis of input variables describing working conditions, such as employment status, occupation details, seniority in the company, main accident hazards, physical demands, psychosocial possibilities, work rhythm determining factors, fit between working hours and family or social commitments, and recent workplace risk assessment. The study conducted by Suárez Sánchez et al. (2011) showed inter-alia that the SVM technique was able on the basis of data on the working conditions to indicate those workers who had suffered from an accident at work in the last year and those who had not been subject to any accidents in the past.

Another example of using machine learning to predict accidents at work was presented by Tixier et al. (2016), who applied Random Forest and Stochastic Gradient Tree Boosting algorithms to a data set of binary attributes and categorical safety outcomes extracted from a large pool of textual construction injury reports by means of the highly accurate Natural Language Processing tool. The developed model was able to predict injury type, energy type, and body part affected with the ability exceeding the previous parametric models.

Another example of using big data methods to improve workplace safety in offline mode is the platform developed by Guo et al. (2016), which enables to classify, collect and store data about unsafe behaviours of construction workers. An intelligent video surveillance system and a mobile application were used to collect information on workers' unsafe behaviours. Behaviour-related data could be retrieved with complete semantic information, including identification, time, image, source, location, description, unsafe behaviour type and a possible injury. The introduction of this system on a metro construction site demonstrated that it could effectively analyse semantic information contained in collected images, automatically extract workers' unsafe behaviour and quickly retrieve this information from the dedicated database system.

An example of machine learning applications in the area of OSH at the level of governmental administration is the Risk Group Prediction Tool (RGPT), developed by the Norwegian Labour Inspection Authority to assist labour inspectors in selecting enterprises with regard to workplace risks (Dahl and Starren 2019). This tool covers approximately 230,000 enterprises in Norway and divides them into four groups according to their OSH-related risks. It is assumed that the higher the risk group, the higher the probability that a future inspection of working conditions will detect deviations from OSH regulations in the company. RGPT was built on the basis of predictive modelling with the help of a machine learning algorithm using binary logistic regression analysis, which is a part of the supervised learning algorithms class. With the increasing number of inspections performed, the predictions made by RGPT become gradually more precise, because the algorithm adjusts itself on the basis of feedback (correct or erroneous forecasts) registered in the database containing data on already performed inspections.

6.4.3 EXAMPLES OF MACHINE LEARNING AND BIG DATA ANALYTICS APPLICATIONS TO SUPPORT OSH MANAGEMENT PROCESSES IN REAL OR NEAR-REAL TIME

A review of scientific literature and websites shows that the solutions for OSH process monitoring and management in real or near-real time based on advanced machine learning and big data analytics methods are still often at the stage of conceptual development or pilot testing in laboratory or industrial conditions. However, the rapid development of technologies supporting the Industry 4.0 concepts suggests that these solutions will soon become more mature and find more practical applications in various industry sectors.

The first example of a solution in this class is the system developed by Yang et al. (2016), which automatically detects and registers near-miss falls based upon kinematic data regarding workers captured by wearable inertial measurement units. The detection of near-miss falls is facilitated by analysing the kinematic data of workers by means of a semi-supervised learning algorithm (Support Vector Machine). The system will make it possible to predict risky locations or hazards in the iron-working industry based upon the detected locations of near-miss falls and will provide information that can be used for introducing proactive fall-prevention measures.

Another example is AI-based visual analytics technology that enables real-time video analysis with advanced algorithms and machine learning to verify if workers are using PPE items correctly. The system is offered by the UK-based Cortexica company under the name *AI-PPE Compliant* (Cortexica 2019; Peniak 2019). A similar purpose is that of another system that involves the use of intelligent vision-based analytics and hierarchical support vector machine algorithm for monitoring whether workers are wearing safety helmets at the workplace or not, and, at the same time, identifying colours of the helmets worn by workers (Wu and Zhao 2018). With regard to these examples, the high potential of AI-based video analytics in the area of OSH is underlined by Naylor (2019), who points to the possibility of applying these technologies to the detection of unsafe working practices and precursors of serious accidents in real time, as well as to auto-reporting of near misses, which will make is possible to deal with the current problem of significant incident under-reporting.

The next area where the harnessing of AI technology can significantly improve working conditions is the development of human-machine interfaces (HMI) which would be capable of automatically adapting to the current state of health and mental workload of operators controlling complex machinery or automated production systems. An example of this is the cyber-physical system supporting human-robot interaction and training to ensure safe human-robot collaboration in manufacturing environments (Tsiakas et al. 2017). The operator's behaviour and mental state while interacting with a robot can be diagnosed using a variety of visual, physiological and linguistics modalities such as facial expressions, gestures, heart rate, temperature, language syntactic relations, etc. Based on this data and the algorithms of Interactive Reinforcement Learning the robot will be able to learn and adapt its collaboration policy to the needs and behaviours of individual operators.

Another promising direction for the development of AI systems to support OSH management processes is the introduction of non-invasive techniques of electroencephalography (EEG) to monitor stress and cognitive load of workers in real time. One of the solutions that applies EEG is a system diagnosing increased stress levels among construction workers (Jebelli et al. 2018). This system is based on the use of a wearable EEG headset (Emotiv EPOC+) and a machine learning algorithm called Gaussian Support Vector Machine. In turn, Neu et al. (2019) developed a concept of what is called the Cognitive Work Protection system, in which they proposed the use of EEG technique to measure workers' cognitive conditions, e.g. the stress level or the ability to concentrate, in order to optimise the interaction with robots and machines in real time. The objective is to reduce the number of accidents at work that involve co-operation between humans and robots and to promote the physical and mental health of workers.

Yet another approach to using AI methods to improve working conditions is the system currently being conceptualised and developed within the EU-funded Ageing@work project. The main objective is to support ageing workers at home and at work so that they can become actively involved in working life for longer. The leading concept is the creation of a flexible working environment that will self-adapt to workers' changing needs taking into account ergonomic principles (Giakoumis et al. 2019). To achieve this, the project consortium intends to develop, inter alia, the novel Ambient Virtual Coach, which will consist of an empathic avatar for the provision of subtle notifications, an adaptive Visual Analytics–based personal dashboard, and a reward-based motivation system targeting positive and balanced worker behaviour at work and in the home environment.

6.4.4 THE CONCEPT OF OSH MANAGEMENT BASED ON SMART PPE AND WEARABLE TECHNOLOGIES COMBINED WITH THE USE OF IOT AND BIG DATA ANALYTICS

As demonstrated in previous sections, recent developments in smart PPE systems, workplace wearables, IoT industrial networks and industrial applications of machine learning and big data analytics, as well as the growing number of other smart machines and devices being introduced into digitally transformed industries and other business sectors, provide the basis for proposing new approaches to OSH management that could be better suited to digital work environments. These approaches involve the use of a number of new functions, most of which will be delivered automatically by supporting technologies.

These functions are linked to the implementation of various processes and components of the OSH management system and, given their importance and relevance to the achievement of OSH objectives established in the company, it is proposed to divide them and assign roughly to the following three levels: basic, diagnostic, and strategic. The application of these functions is discussed below, taking into account the proposed division.

The basic level considers first of all a new role and importance of smart PPE systems in preventing and reducing OSH risks, while taking into account the relevant priorities for action resulting from the risk control hierarchy, as discussed in section 5.3.4. In addition, the basic functions provided by smart PPE and workplace wearables may also include automatic detection of accidents and near misses, as well as subsequent notification of supervisors, safety managers and/or emergency services on such events.

The next, the diagnostic level, covers other functions that are not directly related to real-time assessment and reduction of OSH risks, but are useful for ongoing monitoring of other processes and making appropriate managerial decisions. Increasingly advanced smart networked PPE systems will consist of an increasing number of sensors that will generate increasing amounts of contextual data in real time. These large datasets, multiplied by the increasing number of workers wearing various types of smart PPE, need to be subject to advanced machine learning and big data analytics in order to be thoroughly analysed, and the resulting conclusions implemented effectively to manage OSH. These methods are not yet fully used in this area, but they have great potential, especially in terms of diagnostic functions, which are needed to monitor the effectiveness of OSH management processes in terms of their compliance with established objectives, enable prediction of the effects of these processes on the state of OSH, and thus support managers in making appropriate decisions. Moreover, data on the effectiveness of individual OSH management processes may be used to better define, select and use key performance indicators (KPIs) that can be applied in this area.

And last but not least, IoT technologies combined with predictive data analytics are increasingly being used for what is known as predictive maintenance to optimise the use of machinery and equipment by eliminating failures and optimising maintenance scheduling based on the measurement of performance factors in real time. Cloud-based services that offer data analysis for predictive maintenance are now available and could also be also applied to smart PPE systems or other intelligent safety-related devices, especially when their proper functioning is essential for the safety and health of workers.

The highest strategic level includes functions enabling the acquisition of knowledge that can be useful for strategic planning and shaping of OSH policy in an enterprise on a longer time scale. For example, using context-based data analytics services with respect to workers' physiological parameters measured by means of smart PPE and wearables systems enables monitoring of workers' health and provides information on individual workers or groups of workers who are more vulnerable or who need additional support or special protection. It is also possible to detect potential unsafe behaviour of workers (e.g. by comparing current data with data patterns obtained during accidents or near misses), and identify high-risk zones by geographical location of areas with higher rates of incidents (such as slips, falls, injuries, etc.) or with more frequent unsafe behaviour of workers.

The use of big data analytics methods can also provide many other useful insights and predictions that will improve OSH management at the strategic level. For example, correlating data on the current exposure of workers to different risk factors with the medical data on the health status of workers will help to assess the actual

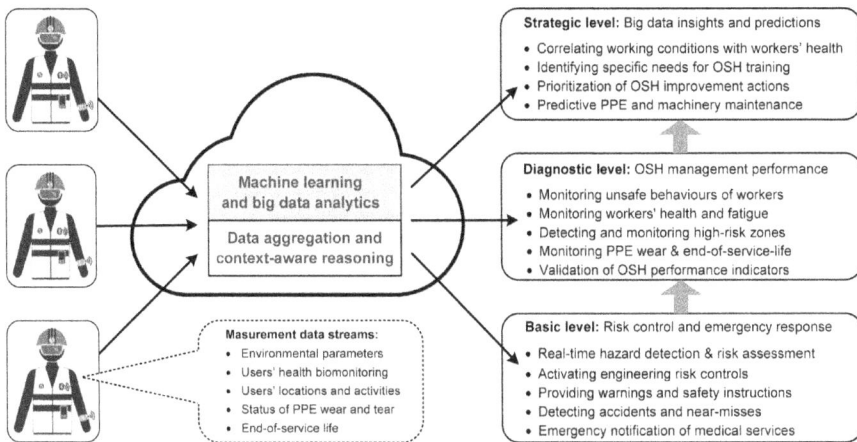

FIGURE 6.6 OSH management functions facilitated by the use of smart PPE and workplace wearables combined IoT technologies, machine learning and big data analytics

impact of working conditions on workers' health and thus, based on the relationships revealed, to apply the most appropriate preventive measures tailored to the needs of vulnerable groups of workers. Similarly, the conclusions derived from big data analytics may also allow the diagnosis of the general level of safety culture across the enterprise or within individual departments or groups of workers, which in turn may lead to a better identification of workers' needs with regard to OSH training, or with regard to their involvement and participation in OSH-related activities.

The above discussion on new functions implemented by smart PPE and wearables combined with the use of IoT and big data analytics, as well as their assignment to different levels of OSH management, can be summarised in the form of a graphical model, which is presented in Figure 6.6.

Most of the advanced functions of OSH management that are based on the use of smart PPE and wearables combined with the use of IoT technologies and big data analytics are already being introduced in some workplace applications, or are still at the stage of development or pilot tests. Some other features are still at the conceptualisation stage, but one can optimistically assume that as the mentioned enabling technologies are being rapidly developed, these new functions will probably be available for introduction into practice in the near future.

6.5 DATA PROTECTION AND CYBER-SECURITY ISSUES WITH REGARD TO THE USE OF SMART PPE AND WEARABLES IN THE WORKPLACE

6.5.1 Defining Workers' Personal Data

As mentioned earlier in this chapter, the dynamic development of enabling technologies in the fields of smart PPE, workplace wearables and the IoT networks has allowed the development and implementation of a number of new useful functions, some of which are dedicated to monitoring the working environment, while other deal with monitoring parameters related to workers' psycho-physical conditions, their effectiveness and performance efficiency and the context of work-related activities. The data obtained and processed by the monitoring functions in relation to workers can be roughly divided into the following three categories:

1) Biometric data concerning worker's health status (i.e. medical data), which may include heart rate (HR), respiration rate, body temperature, blood pressure, oxygen consumption, galvanic skin response (GSR), stress hormones, ECG and EEG data, etc.;
2) User identification, location and interaction data, i.e. data allowing to determine the geographical position of individual user in specific places or in relation to high-risk zones, time spent in specific rooms, zones and spaces, and worker's behaviour based on data concerning contacts with other workers;
3) Worker performance data, which may include energy expenditure, time spent on performing individual tasks, duration of rest breaks, worker' cognitive pattern, etc.

As can be seen from this, most of the types of this data constitute personal data and other sensitive data, which, due to legal requirements, privacy or ethical aspects, should be subject to special attention and protection at all stages of collection, processing and use. The workers' personal data can be defined generally as "any information related to an identified or identifiable worker" (ILO 1997). Similar broad definition was adopted by the Organisation for Economic Co-operation and Development (OECD 2013).

The public concern about the amount of private data that are digitally collected and processed for various business purposes has been growing in recent years, particularly in the EU countries, resulting finally in the entry into force of a new European law in 2018, namely the General Data Protection Regulation (European Union 2016b), commonly known as GDPR. A definition of personal data adopted in this regulation seems to be more adequate and applicable to the discussion on workers' privacy issues in the context of ICT-based monitoring, as it states that personal data is:

any information relating to an identified or identifiable natural person ("data subject"); an identifiable natural person is one who can be identified, directly or indirectly, in particular by reference to an identifier such as a name, an identification number, location data, an online identifier or to one or more factors specific to the physical, physiological, genetic, mental, economic, cultural or social identity of that natural person.

In other countries of the world the definitions of personal data, as well as respective privacy regulations, vary greatly depending on the level of sensitisation of the societies and the governments to the problem of personal data protection. A particular case of differentiation of legal solutions in this area occurs in the United States, where there is no uniform federal law on the protection of personal data, but these issues are regulated differently at the level of state administration (Hickman et al.

2019). But given global trends and growing concerns that citizens' personal data are not properly protected against unauthorised processing, new regulations are being introduced in successive US states to implement procedures for the protection of personal data similar to those resulting from the GDPR. One of the best-known and recent examples of such state-level initiatives is the California Consumer Privacy Act (CCPA 2018), which became effective on January 1, 2020. The CCPA defines personal information as any information that identifies, relates to, describes, is capable of being associated with, or could reasonably be linked, directly or indirectly, with a particular consumer or household. Then, the definition provides a list of types of information to which the CCPA applies that is longer and more detailed compared to GDPR as it includes inter alia: identifiers such as a real name, alias, postal address, unique personal identifier, email address, account name, social security number, driver's licence number, passport number, biometric data, Internet or other electronic network activity information, geolocation data, education history, professional or employment-related information, purchase history, and inferences drawn from any of the aforementioned information to create a profile about a consumer reflecting the consumer's preferences, characteristics, psychological trends, preferences, predispositions, behaviour, attitudes, intelligence, abilities, and aptitudes.

6.5.2 LEGAL PROCEDURES OF WORKERS' PERSONAL DATA PROTECTION – AN EXAMPLE OF THE EU APPROACH

The GDPR distinguishes what is known as special category data, which includes personal data concerning health understood as data related to the physical or mental health of a natural person which reveals information about his/her health status. Then, the most essential clause of the GDPR stipulates that the processing of special category data should be generally prohibited, which means that this restriction applies also to workers' data that may reveal their health status. At the same time, other GDPR provisions state that processing of personal data concerning health can be allowed if the data subject has given explicit consent to the data processing, or processing is necessary for the purposes of preventive or occupational medicine, or for the assessment of the working capacity of the worker. It follows from the above that data concerning physiological parameters of workers determining their health condition, which can be collected and processed by systems based on smart PPE and workplace wearables, are subject to strict legal protection. Thus, designers and suppliers of these systems should be aware of these limitations and take them into account already at the stage of design and testing, as well as at the stage of final installation and putting into operation in a given workplace.

Moreover, account should also be taken of the fact that all personal data collected at the workplace can be used for worker profiling. GDPR defines profiling as:

> any form of automated processing of personal data consisting of the use of personal data to evaluate certain personal aspects relating to a natural person, in particular to analyse or predict aspects concerning that natural person's performance at work, economic situation, health, personal preferences, interests, reliability, behaviour, location or movements.

It follows from this definition that many functions useful for OSH management performed by smart PPE and workplace wearables systems on the basis of context data analysis, such as health status data, user location, behaviour patterns, work performance, etc., should be regarded as specific cases of profiling. GDPR states also that "the data subject shall have the right not to be subject to a decision based solely on automated processing, including profiling, which produces legal effects concerning him or her or similarly significantly affects him or her", and subsequently, fortunately, allows a derogation from this rule if data subject has given an explicit "consent" to be subject to a decision based on profiling. The term "consent" is to be understood as "any freely given, specific, informed and unambiguous indication of the data subject's wishes by which he or she, by a statement or by a clear affirmative action, signifies agreement to the processing of personal data relating to him or her". A typical form of such a consent is a written statement delivered by the data subject, preferably signed. Electronic forms of explicit consent such as online forms, emails, scanned documents with a person's signature and electronic signatures are also permissible and may be used (Mulder 2019).

The analysis of the above provisions indicates that in order for the systems based on smart PPE and workplace wearables, which automatically process personal data for one of the purposes indicated in the definition of profiling, to comply with legal requirements and be accepted by workers, a number of technical and organisational conditions must be met to guarantee a high level of protection of personal data processed. Moreover, a very important aspect of effective implementation of these systems is the prior consent of workers to the use these data for decisions based on profiling, which should be appropriately confirmed by a documented, understandable to the worker and consciously expressed consent.

In order to ensure appropriate technical and organisational measures to adequately protect personal data, the GDPR requires data processing system administrators (referred to in GDPR as "data controllers") to apply two guiding principles: (1) data protection by design; and (2) data protection by default. The principle "data protection by design" means the use of appropriate technical and organisational measures, such as pseudonymisation and data minimisation, in an effective manner, and the integration of the necessary safeguards into the data processing. Pseudonymisation is defined as:

> processing of personal data in such a manner that the personal data can no longer be attributed to a specific data subject without the use of additional information, provided that such additional information is kept separately and is subject to technical and organisational measures to ensure that the personal data are not attributed to an identified or identifiable natural person.

And data minimisation is to be understood as limiting collection and processing of data to what is necessary in relation to the purposes for which it is processed.

On the other hand, the application of technical and organisational measures according to "data protection by default" principle means:

> ensuring that, by default, only personal data which are necessary for each specific purpose of the processing are processed. That obligation applies to the amount of personal data collected, the extent of their processing, the period of their storage and their

accessibility. In particular, such measures shall ensure that by default personal data are not made accessible without the individual's intervention to an indefinite number of natural persons.

According to this principle, the preferences and needs of the persons whose data is to be processed should be the starting point for any processing of personal data. The controller of a data processing system should always (by default) ensure the highest possible level of protection of personal data for all users, while allowing the possibility for individual users to change these initial settings according to their preferences.

A review of the scientific literature on the development of OSH-related systems based on smart PPE, wearables and IoT technologies shows still a relatively low level of awareness of legal obligations to guarantee the protection of workers' personal data that will be collected and processed in these systems. New regulations on the protection of personal data have been introduced in the EU, the United States and in other countries quite recently, so the awareness of these regulations, their interpretation in relation to specific technological solutions, and the subsequent transfer of this knowledge into practice will take some time. However, it can be assumed that the new solutions of the systems discussed here, which are already being developed or will be developed in the future, will be more privacy-friendly and more mature in terms of ensuring the protection of personal data of the future users.

Various guidelines and good practices can contribute to further progress in this area, with the aim of helping both designers and system providers, as well as employers and workers, to understand the requirements of personal data protection and to translate them into concrete technological solutions that take into account the working environment conditions and expectations of all the stakeholders. An example of such guidelines is the UNI Global Union (2017) publication containing 10 principles for the protection of workers' privacy and personal data in the context of the increasing scale of digital processing of such data. Another example is the guide of the European Federation of Public Service Unions (EPSU 2019) that examines the GDPR from three different perspectives: the impact on workers, the impact on public services and the impact on trade unions. Whereas examples of guidelines that are particularly relevant to broadening knowledge on the implementation of personal data protection regulations with regard to ICT-based systems are the publications of the EU Agency for Network and Information Security (ENISA), including in particular recommendations for data pseudonymisation (ENISA 2018a), and the recommendations for the implementation of the "data protection by default" principle (ENISA 2019a).

6.5.3 CYBER-SECURITY ISSUES WITH REGARD TO ICT-BASED SYSTEMS CONSISTING OF SMART PPE AND WORKPLACE WEARABLES

As information and communication technologies, smart distributed systems and overall connectivity increase and become more widespread in many areas of life and economy sectors, cyber-security issues become increasingly important as systems become more vulnerable to cyber-attacks and the number of reported attacks increases rapidly. Cyber-security issues are also important in the manufacturing sector, to ensure the proper and uninterrupted functioning of the intelligent manufacturing systems covered by the industrial IoT networks, including those that incorporate smart PPE and workplace wearables. In the latter case, the inclusion of cyber-security issues does not have only an operational aspect but also is subject to the requirements of protecting privacy of workers covered by these systems.

The concepts of privacy and personal data protection and cyber-security are inter-related, but not synonymous. Privacy protection is basically a legal and ethical concept and consists in building awareness of the risks to privacy, ensuring individual control over the collection and processing of personal data and controlling the use and dissemination of personal information to any entity outside the sphere of personal control of a given data subject (see section 6.5.1). In turn, cyber-security, defined as "measures taken to protect a computer or computer system (as on the Internet) against unauthorized access or attack" (Merriam-Webster 2020b) belongs to the sphere of technology and management and concerns activities aimed at preventing attacks on various types of data, including data needed for the efficient functioning of the system, data affecting the safety of users, and other personal sensitive data related to users' privacy.

So far, there have been no reports on direct cyber-attacks on IoT-based systems consisting of smart PPE or workplace wearables, but many cyber-threats and attacks on IoT devices and systems, as well as attacks on ICT-based industrial installations and devices, have already been reported (Dunlap 2017; McLaughlin et al. 2016; Kaspersky lab ICS CERT 2019). In the context of the direct relationship between cyber-attacks and occupational safety and health, it is worth considering the case of cyber-attacks on industrial installations using malware called TRISIS, which is tailored to industrial Safety Instrumented Systems, i.e. systems designed to keep an industrial system or a plant in a safe state when abnormal conditions occur (Dragos 2017; Buntz 2019).

From the point of view of the potential impact on OSH cyber-risks can be analysed taking into account the division of these risks into three categories (More and Allnutt 2017):

Operational: the disruption of business operations caused by the loss or interruption of electronic systems and ICT-based networks.

Informational: the loss, unauthorised accessing, destruction, or other unintended use of electronic information and collected data.

Physical: physical damage of the equipment, machinery, installations etc. or unexpected physical events caused by actions in the cyber domain.

In order to illustrate the potential consequences of cyber-attacks against ICT-based systems comprising smart PPE and workplace wearables, the following three hypothetical scenarios are presented:

1. *Operational cyber-risk*: a cyber-attack may disable functions of a health monitoring system of lone workers and may lead to leaving them unsupported in hazardous conditions (based on More and Allnutt (2017)).
2. *Informational cyber-risk*: stealing workers' health information by a hacker to use it for blackmailing and extortion or to sell for marketing purposes.

3. *Physical cyber-risk*: hacking worker's smart glasses (i.e. glasses containing augmented reality modules) that will allow an attacker to take control of the system and display false instructions; this in turn may result in incorrect actions, leading to damage to machinery, installations or the entire industrial system.

The general mechanism of the impact of possible cyber-attack on the protective performance of smart PPE system, and thus on the safety and health of workers wearing components of such system, is shown in Figure 6.7. The concept of subsequent mechanism stages was inspired by the demonstration of cyber-security aspects of safety-related control systems of machinery, which is included in the IEC Technical Report no. 63074 (IEC 2019b). The presented mechanism not only involves traditional aspects of OSH, but also shows another, no less important negative consequence of such a cyber-attack, which may be loss of privacy of a user or another malicious use of his/her sensitive personal data.

To date (January 2020), no guidelines or standards have yet been developed concerning the inclusion of cyber-security aspects, which would refer exclusively to the design of IoT-based systems consisting of smart PPE or workplace wearables. However, these systems are principally based on the same design concepts, connected technologies, hardware components and computing platforms as other industrial IoT systems used in the industry. Therefore, in order to prevent potential cyber-attacks on smart PPE-based systems and to limit the above described consequences, it is advisable to selectively follow cyber-security recommendations and good practices provided in normative documents relating to the entire class of IoT systems, and in particular to industrial applications of such systems. First of all, ENISA publications which contain cyber-security good practices for IoT in smart factories (ENISA 2018b) and cyber-security recommendations in the context of Industry 4.0 (ENISA 2019b) should be mentioned. In addition, relevant information for application in this field is contained in normative documents which are already developed or under development by such standardisation bodies and committees as the Technical Committee ISO/IEC JTC 1/SC 27 on Information Security, Cybersecurity and Privacy Protection (ISO 2019), Technical Committee CYBER of the European Telecommunications Standards Institute (ETSI 2019), and Technical Committee CEN/CLC/JTC 13 on Cybersecurity and Data Protection (CEN 2019). In line with this, the UL 2900-1 standard on software cyber-security for network-connectable products (UL 2017), which is applicable in the United States and Canada, can also be regarded as valuable in context under consideration.

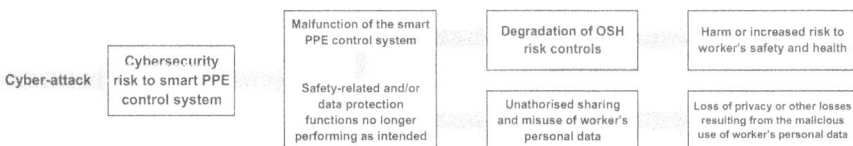

FIGURE 6.7   The mechanism of a possible cyber-attack impact on the safety and health and privacy of workers wearing smart PPE components

A review of the aforementioned publications and the analysis of other literature on cyber-security issues enables to formulate a number of the general recommendations relating to the technical sphere of cyber-security, which can be useful at the current stage of development of the systems in question:

- applying user authentication and authorisation (including biometric techniques), i.e. verifying the identity of a user (but preferably ensuring anonymity at the same time) or a device as a prerequisite for granting the access to system functions;
- introducing secure boot mechanisms to monitor any changes in hardware and software of the system – if an unauthorised change is detected, the device should alert the user and/or the system administrator;
- using edge computing solutions rather than cloud-based systems to limit amount of data subject to long-distance transmission and online processing;
- if cloud-based services need to be used, selecting those that ensure the highest level of data protection and resistance to cyber-attacks;
- using efficient cyber-security protocols and encrypting algorithms, particularly in case of the transmission of sensitive personal data; and
- continuous anti-malware testing and using anti-virus tools to prevent, detect, and remove any malicious software introduced into the system.

6.6 SUMMARY

The concepts and practical examples smart PPE systems and workplace wearables connected through the networks of industrial IoT, which are presented and discussed in this chapter, clearly show that these solutions have excessive potential to support the basic OSH risk management process, and to provide a vast amount of data to be used in a broader OSH management context by searching for useful actionable insights based on big data analytics. But, at the same time, these solutions bring a number of constraints and other challenges, mainly of an ethical and legal nature, which should be taken into account in the further development of these technologies so as to ensure their effectiveness, usability and acceptability by workers, employers and other stakeholders.

First of all, the designers and providers of these new systems should be very respectful of the protection of personal data collected and processed in these systems, particularly taking into account that medical data measured in the workplace, and then processed to support OSH management belong to sensitive personal data that is owned by the workers. Thus, workers, as the end users, have every right to require that their data is rigorously protected by both legal regulations and necessary technologies, and that the insight produced as a result of the data analytics will be used only and exclusively for the benefit of workers.

In the context of the protection of personal data collected and processed within ICT-based networks consisting of smart PPE systems and workplace wearables, it is particularly important to appropriately address cyber-security aspects, in particular in the design and implementation phases. Industrial IoT systems covering smart PPE and wearables can nowadays be targeted by various cyber-attacks, as are other industrial IoT networks, including those systems that are responsible for ensuring the safety and health in the workplace. Therefore, future studies aimed at the creation and application of novel technologies for smart PPE and other workplace wearables should include the elaboration of appropriate methods to assess the vulnerability of these systems to the growing cyber-risks, as well as the development of respective technological and organisational measures to effectively protect these systems against these risks. Achieving a sufficiently high level of protection of these systems against cyber-risks will be an essential prerequisite for the further dissemination of these systems worldwide, and for their successful acceptance by the end users.

REFERENCES

Abtahi F., Diaz-Olivazrez A., Forsman M. et al., 2017. Big data & wearable sensors ensuring safety and Health @Work. In: *GLOBAL HEALTH 2017, the Sixth International Conference on Global Health Challenges*. http://kth.diva-portal.org/smash/get/div a2:1195845/FULLTEXT01.pdf (accessed August 6, 2019).

Ajayi A.O., Oyedele L., Davila Delgado J.M., Akanbi L., Bilal M., Akinade O., Olawale O. 2018. Big data platform for health and safety accident prediction. *World Journal of Science, Technology and Sustainable Development* 16(1):2–21.

Awolusi I., Marks E., Hallowell M., 2018. Wearable technology for personalized construction safety monitoring and trending: Review of applicable devices. *Automation in Construction* 85:96–106. https://www.sciencedirect.com/science/article/pii/ S092658 0517309184 (accessed July 20, 2019).

Ayodele T.O., 2010. Types of machine learning algorithms. In: *New Advances in Machine Learning*, Yagang Zhang (Ed.), Intech Open. http://cdn.intechweb.org/pdfs/10694.pdf (accessed August 5, 2019).

Bagula A.B., Osunmakinde I., Zennaro M., 2010. On the relevance of using bayesian belief networks in wireless sensor networks situation recognition. *Sensors* 10(12):11001–11020. http://wireless.ictp.it/Papers/sensors-10-11001.pdf (accessed August 5, 2019).

Barata J., da Cunha P.R., 2019. Safety is the new black: The increasing role of wearables in Occupational Health and safety in construction. In: *Lecture Notes in Business Information Processing, Springer*, W. Abramowicz, R. Corchuelo (Eds.), vol. 353, 526–537. Business Information Systems. BIS 2019.

Bartkowiak G., Dąbrowska A., Marszałek A., 2013. Analysis of thermoregulation properties of PCM garments on the basis of ergonomic tests. *Textile Research Journal* 83(2):148–159.

Bettini C., Brdiczka O., Henricksen K., Indulska J., Nicklas D., Ranganathan A., Riboni D., 2010. A survey of context modelling and reasoning techniques. *Pervasive and Mobile Computing* 6(2):161–180.

Bruckner D., Velik R., 2010. Behavior learning in dwelling environments with hidden markov models. *IEEE Transactions on Industrial Electronics* 57(11):3653–3660.

Buntz B., 2019. Trisis malware discovered at additional Industrial Facility. IoT. *World Today*. https://www.iotworldtoday.com/2019/04/11/trisis-malware-discovered-at-additional-industrial-facility/ (accessed August 8, 2019).

CCPA, 2018. 1.81.5. California consumer privacy act of 2018 (1798.100 – 1798.199) (2018), CAL. CIV. CODE. https://leginfo.legislature.ca.gov/faces/codes_displayText.xhtml? division=3.&part=4.&lawCode=CIV&title=1.81.5 (accessed January 7, 2020).

CEN (European Committee for Standardization), 2011. *CEN/TR 16298 Textiles and Textile Products – Smart Textiles – Definitions, Categorisation, Applications and Standardization Needs*. Brussels, Belgium: European Committee for Standardization.

CEN (European Committee for Standardization), 2019. CEN/CLC/JTC 13 – Cybersecurity and data protection. European Committee for Standardization. https://standards.ce n.eu/dyn/www/f?p=204:7:0::::FSP_ORG_ID:2307986&cs=1E7D8757573B5975ED28 7A29293A34D6B (accessed January 7, 2020).

Chahuara P., Portet F., Vacher M., 2013. Making context aware decision from uncertain information in a smart home: A Markov logic network approach: Ambient intelligence. *Lecture Notes in Computer Science, Springer* 8309:78–93.

Choi B., Hwang S., Lee S.H., 2017. What drives construction workers' acceptance of wearable technologies in the workplace?: Indoor localization and wearable health devices for occupational safety and health. *Automation in Construction* 84:31–41. https://ww w.sciencedirect.com/science/article/pii/S0926580517307215 (accessed July 30, 2019).

Choo K.H., Tong J.C., Zhang L., 2004. Recent applications of hidden markov models in computational biology. *Genomics, Proteomics and Bioinformatics* 2(2):84–96. https://ww w.ncbi.nlm.nih.gov/pubmed/15629048 (accessed July 30, 2019).

Cortexica, 2019. Overcoming the barriers of PPE compliance with the use of artificial intelligence. AI-PPE compliant. White paper. London, UK: Cortexica. https://www.cortexic a.com/ai-ppe-compliance-whitepaper (accessed August 6, 2019).

Dahl Ø., Starren A., 2019. The future role of big data and machine learning in health and safety inspection efficiency. Discussion paper. Bilbao, Spain: European Agency for Safety and Health at Work (EU-OSHA).

Dai F., Olorunfemi A., 2018. *Holographic Visual Interaction and Remote Collaboration in Construction Safety and Health.* Morgantown: West Virginia University, USA. https:// www.cpwr.com/sites/default/files/publications/SS2018-visual-interaction-remote-col laboration.pdf (accessed August 8, 2019).

Dempster A.P., 1966. New methods for reasoning towards posterior distributions based on sample data. *The Annals of Mathematical Statistics* 37(2):355–374.

Dey A.K., 2001. Understanding and using context. *Personal and Ubiquitous Computing* 5(1):4–7.

Dey A., 2016. Machine learning algorithms: A review. *International Journal of Computer Science and Information Technologies* 7(3):1174–1179. http://ijcsit.com/docs/Volume% 207/vol7issue3/ijcsit2016070332.pdf (accessed August 5, 2019).

Dolez P., Decaens J., Buns T., Lachapelle D., Vermeersch O., Mlynarek J., 2018. Analyse du potential d'application des textiles intelligents en santé et en sécurite au travail. Rapports Scientifiques R-1029. Montréal, Québec, Canada: Institut de recherché Robert-Sauvé en santé et en sécurite du travail (IRSST). https://www.irsst.qc.ca/media/ documents/PubIRSST/R-1029.pdf?v=2019-10-17 (accessed August 5, 2019).

Dragos, 2017. TRISIS malware: Analysis of safety system targeted malware. Hanover, MD: Dragos Inc. https://dragos.com/wp-content/uploads/TRISIS-01.pdf (accessed August 8, 2019).

Dunlap T., 2017. The 5 worst examples of IoT hacking and vulnerabilities in recorded history. https://www.iotforall.com/5-worst-iot-hacking-vulnerabilities/ (accessed August 8, 2019).

Elokon, 2019. New: Smart safety vest for ELOshield: Innovative personnel detection – No additional module needed. Elokon GmbH. https://www.elokon.com/en-EN/newsroom/ details/smart-safety-vest-for-eloshield.html (accessed July 30, 2019).

ENISA (European Union Agency For Network and Information Security), 2018a. Recommendations on shaping technology according to GDPR provisions: An overview on data pseudonymisation. European Union Agency For Network and Information Security. https://www.enisa.europa.eu/publications/recommendations-on-shaping-te chnology-according-to-gdpr-provisions (accessed August 8, 2019).

ENISA (European Union Agency For Network and Information Security), 2018b. Good practices for security of internet of things in the context of smart manufacturing. European Union Agency For Network and Information Security. https://www.enisa.europa.eu/pu blications/good-practices-for-security-of-iot (accessed August 9, 2019).

ENISA (European Union Agency For Network and Information Security), 2019a. Recommendations on shaping technology according to GDPR provisions – Exploring the notion of data protection by default. European Union Agency For Network and Information Security. https://www.enisa.europa.eu/publications/recommendations-on-shaping-technology-according-to-gdpr-provisions-part-2 (accessed August 8, 2019).

ENISA (European Union Agency For Network and Information Security), 2019b. Industry 4.0 cybersecurity: Challenges & recommendations. European Union Agency For Network and Information Security. https://www.enisa.europa.eu/publications/industry-4-0-cybersecurity-challenges-and-recommendations (accessed August 9, 2019).

EPSU (European Federation of Public Service Unions), 2019. The General Data Protection Regulation (GDPR): An EPSU briefing. Brussels, Belgium: European Federation of Public Service Unions. https://www.jpgu.org/sites/default/files/article/files/GDPR_FINAL_EPSU.pdf (accessed August 8, 2019).

ETSI (European Telecommunications Standards Institute), 2019. Technical Committee (TC) CYBER (Cyber Security). European Telecommunications Standards Institute. https://www.etsi.org/committee/cyber?jjj=1565335091816 (accessed August 9, 2019).

European Union, 2016a. Regulation (EU) 2016/425 of the European Parliament and of the Council of 9 March 2016 on personal protective equipment and repealing Council Directive 89/686/EEC. *Official Journal of the European Union* L 81,51-98.

European Union, 2016b. Regulation (EU) 2016/679 of the European Parliament and of the Council of 27 April 2016 on the protection of natural persons with regard to the processing of personal data and on the free movement of such data, and repealing Directive 95/46/EC (General Data Protection Regulation). *Official Journal of the European Union* L119, 1-88.

Fink G.A., 2014. *Markov Models for Pattern Recognition: From Theory to Applications.* 2nd edition. London, UK: Springer-Verlag.

Fugini M., Teimourikia M., 2014. Risks in smart environments and adaptive access controls. In: *Proceedings of the 4th International Conference on Innovative Computing Technology (INTECH)*, 13–15 August 2014, Luton, UK, 97–102.

Gales M., Young S., 2007. The application of hidden Markov models in speech recognition. *Foundations and Trends in Signal Processing* 1(3):195–304.

Gandomi A., Haider M., 2015. Beyond the hype: Big data concepts, methods, and analytics. *International Journal of Information Management* 35(2):137–144. https://www.sciencedirect.com/science/article/pii/S0268401214001066#bib0055 (accessed August 5, 2019).

García-Herrero S., Mariscal M.A., García-Rodríguez J., Ritzel D.O., 2012. Working conditions, psychological/physical symptoms and occupational accidents: Bayesian network models. *Safety Science* 50(9):1760–1774.

Gartner IT Glossary, 2020a. Big Data. https://www.gartner.com/it-glossary/big-data (accessed January 7, 2020).

Gartner IT Glossary, 2020b. Data mining. https://www.gartner.com/it-glossary/data-mining (accessed January 7, 2020).

Ghani S., ElBialy E.M.A.A., Bakochristou F., Gamaledin S.M.A., Rashwanet M.M., 2017. The effect of forced convection and PCM on helmets' thermal performance in hot and arid environments. *Applied Thermal Engineering* 111:624–637. https://www.sciencedirect.com/science/article/pii/S1359431116319056 (accessed July 30, 2019).

Giakoumis D., Votis K., Altsitsiadis E., Segkouli S., Paliokas J., 2019. Smart, personalized and adaptive ICT solutions for active, healthy and productive ageing with enhanced workability. *PETRA'19 Proceedings of the 12th ACM International Conference on PErvasive Technologies Related to Assistive Environments*, 5–7 June 2019, Rhodes, Greece, 442–447.

Giretti A., Carbonari A., Naticchia B., DeGrassi M., 2009. Design and first development of an automated real-time safety management system for construction sites. *Journal of Civil Engineering and Management* 15(4):325–336.

Guo Y., Ding L.Y., Luo H.B., Jiang X.Y., 2016. A Big-Data-based platform of workers' behavior: Observations from the field. *Accident Analysis and Prevention* 93:299–309. https://www.sciencedirect.com/science/article/pii/S0001457515300816 (accessed August 5, 2019).

Haghighi P.D., Krishnaswamy S., Zaslavsky A., Gaberet M.M., 2008. Reasoning about context in uncertain pervasive computing environments: Smart sensing and context. *Lecture Notes in Computer Science, Springer* 5279:112–125.

Hickman T., Gabel D., Goetz M., Chabinksy S., Pittmanet F.P., 2019. *The International Comparative Legal Guide to: Data Protection 2019*. 6th edition. A practical cross-border insight into data protection law. London, UK: Global Legal Group. https://iapp.org/media/pdf/resource_center/comparative_legal_guide_2019.pdf (accessed January 5, 2020).

Hong J., Suh E., Kim S.-J., 2009. Context-aware systems: A literature review and classification. *Expert Systems with Applications* 36(4):8509–8522.

IEC (International Electrotechnical Commission), 2019a. Wearable electronic devices and technologies – parts 101–1: Terminology. Committee Draft IEC 63203-101-1 ED1. Geneva, Switzerland: International Electrotechnical Commission.

IEC (International Electrotechnical Commission), 2019b. Safety of machinery – security aspects related to functional safety of safety-related control systems. IEC TR 63074:2019. Geneva, Switzerland: International Electrotechnical Commission.

ILO (International Labour Organization), 1997. Protection of workers' personal data. An ILO code of practice. Geneva, Switzerland: International Labour Office. https://www.ilo.org/public/libdoc/ilo/1997/97B09_118_engl.pdf (accessed August 7, 2019).

ILO (International Labour Organisation), 2001. Guidelines on occupational safety and health management systems (ILO-OSH 2001). Geneva, Switzerland: International Labour Office. https://www.ilo.org/global/publications/ilo-bookstore/order-online/books/WCMS_PUBL_9221116344_EN/lang--en/index.htm (accessed June 29, 2019).

Investopedia, 2020a. Wearable technology. https://www.investopedia.com/terms/w/wearable-technology.asp (accessed January 7, 2020).

Investopedia, 2020b. Deep learning. https://www.investopedia.com/terms/d/deep-learning.asp (accessed January 7, 2020).

ISO (International Organization for Standardization), 2018. ISO 45001:2018, Occupational health and safety management systems – Requirements with guidance for use. Geneva, Switzerland: International Organization for Standardization.

ISO (International Organization for Standardization), 2019. ISO/IEC JTC 1/SC 27 Information security, cybersecurity and privacy protection. International Organization for Standardization. https://www.iso.org/committee/45306.html (accessed August 9, 2019).

Jebelli H., Hwang S., Lee S.H., 2018. EEG-based workers' stress recognition at construction sites. *Automation in Construction* 93:315–324. https://www.sciencedirect.com/science/article/pii/S092658051830013X (accessed August 7, 2019).

Kagermann H., Lukas W.-D., Wahlster W., 2011. Industrie 4.0: Mit dem Internet der Dinge auf dem Weg zur 4. Industriellen revolution. VDI Nachrichten. http://www.wolfgang-wahlster.de/wordpress/wp-content/uploads/Industrie_4_0_Mit_dem_Internet_der_Dinge_auf_dem_Weg_zur_vierten_industriellen_Revolution_2.pdf (accessed July 30, 2019).

Kaspersky lab ICS CERT, 2019. Threat landscape for industrial automation systems. H2 2018, Kaspersky Lab Industrial Control Systems Cyber Emergency Response Team. https://ics-cert.kaspersky.com/media/KL_ICS_CERT_H2_2018_REPORT_EN.pdf (accessed August 8, 2019).

Kelleher J.D., Mac Namee B., D'Arcy A., 2015. *Fundamentals of Machine Learning for Predictive Data Analytics: Algorithms, Worked Examples, and Case Studies*. Cambridge, MA; London, UK: The MIT Press.

Kenett R.S., 2012. Applications of Bayesian networks. https://papers.ssrn.com/sol3/papers.cfm?abstract_id=2172713 (accessed August 2, 2019).

Khakurel J., Melkas H., Porras J., 2018. Tapping into the wearable device revolution in the work environment: A systematic review. Information Technology & People. https://www.emerald.com/insight/content/doi/10.1108/ITP-03-2017-0076/full/html (accessed July 26, 2019).

Khosrowpour A., Fedorov I., Holynski A., Niebles J., Niebles C., Golparvar-Fard M., 2014. Automated worker activity analysis in indoor environments for direct-work rate improvement from long sequences of RGB-D images. In: *Proceedings of the Construction Research Congress*, 19–21 May 2014, Atlanta, GA, 729–738. http://sip b.sggw.pl/CRC2014/data/papers/9780784413517.075.pdf (accessed August 6, 2019).

Kirchsteiger C., 1999. On the use of probabilistic and deterministic methods in risk analysis. *Journal of Loss Prevention in the Process Industries* 12(5):399–419.

Koshmak G., Linden M., Loutfi A., 2014. Dynamic Bayesian networks for context-aware fall risk assessment. *Sensors* 14(5):9330–9348.

Ku J.-H., Park D.-K., 2013. Developing safety management systems for track workers using smart phone GPS. *International Journal of Control and Automation* 6(5):137–148.

Kubat M., 2017. An introduction to machine learning. Cham, Switzerland: Springer International Publishing AG. https://link.springer.com/book/10.1007%2F978-3-319-63913-0 (accessed August 5, 2019).

Leu S.-S., Chang C.-M., 2013. Bayesian-network-based safety risk assessment for steel construction projects. *Accident Analysis and Prevention* 54:122–133. http://isiarticles.com/bundles/Article/pre/pdf/29205.pdf (accessed July 30, 2019).

Liu X., Wang X., 2017. Development of a smart work site for early warning of heat stress. In: *Proceedings of the 20th International Symposium on Advancement of Construction Management and Real Estate*, Y. Wu, S. Zheng, J. Luo, W. Wang, Z. Mo, L. Shan (Eds.), Singapore: Springer. https://link.springer.com/chapter/10.1007/978-981-10-0855-9_76 (accessed July 30, 2019).

Manuele F., 2005. Risk Assessment & Hierarchies of Control: Their growing importance to the SH&E profession. *Professional Safety* 50(5):33–39. https://aeasseincludes.assp.org/professionalsafety/pastissues/050/05/030505as.pdf (accessed July 30, 2019).

Marchal P., Baudoin J., 2018. General principles of smart personal protection system design. In: *Proceedings of the 9th International Conference on Safety of Industrial Automated Systems (SIAS 2018)*, 1–12 October 2018. Nancy, France, 122–128. http://www.inrs-sias 2018.fr/upload/Proceedings%20SIAS2018.pdf (accessed August 5, 2019).

Mardonova M., Choi Y., 2018. Review of wearable device technology and its applications to the mining industry. *Energies* 11(3):547. https://www.mdpi.com/1996-1073/11/3/547 (accessed July 26, 2019).

Marhavilas P.K., Koulouriotis D., Gemeni V., 2011. Risk analysis and assessment methodologies in the work sites: On a review, classification and comparative study of the scientific literature of the period 2000–2009. *Journal of Loss Prevention in the Process Industries* 24(5):477–523. https://www.sciencedirect.com/science/article/pii/S0950423011000325 (accessed July 29, 2019).

Marhavilas P.K., Koulouriotis D.E., 2012. Developing a new alternative risk assessment framework in the work sites by including a stochastic and a deterministic process: A case study for the Greek Public Electric Power Provider. *Safety Science* 50(3):448–462. https://www.sciencedirect.com/science/article/pii/S0925753511002773 (accessed July 29, 2019).

Mari D., 2014. Big data. The 5 Vs everyone must know. *LinkedIn*. Posted 6 March 2014. https://www.linkedin.com/pulse/20140306073407-64875646-big-data-the-5-vs-everyone-must-know/ (accessed August 5, 2019).

Martin J.E., Rivas T., Matías J.M., Taboada J., Argüelles A., 2009. A Bayesian network analysis of workplace accidents caused by falls from a height. *Safety Science* 50:1760–1774. https://www.sciencedirect.com/science/article/pii/S0925753508000416 (accessed August 5, 2019).

McLaughlin S., Konstantinouz C., Wang X., Davi L., Sadeghi A.R., Maniatakos M., Karri R., 2016. The industrial control systems cyber security landscape. *Proceedings of the IEEE* 104(5). https://ieeexplore.ieee.org/document/7434576 (accessed August 9, 2019).

Merriam-Webster, 2020a. Artificial intelligence. https://www.merriam-webster.com/dictiona ry/artificial%20intelligence (accessed August 2, 2019).

Merriam-Webster, 2020b. Cybersecurity. https://www.merriam-webster.com/dictionary/cy bersecurity (accessed January 6, 2020).

Moore C., Allnutt H., 2017. Cyber risk and the impact on health and safety. SHP Online, https://www.shponline.co.uk/risk/cyber-risk-and-the-impact-on-health-and-safety/ (accessed August 8, 2019).

Mulder T., 2019. Health apps, their privacy policies and the GDPR. *European Journal of Law and Technology* 10(1). http://ejlt.org/article/view/667/8972019 (accessed August 8, 2019).

Naylor S., 2019. Rise in industry 4.0 and industrial AI technologies in workplaces and current and future opportunities for more proactive health and safety: Part 2. https://www.dis coveringsafety.com/blogs/rise-industry-use-ai-technologies-workplaces-and-future -challenges-health-and-safety-part-2 (accessed January 7, 2020).

Necsulescu D.-S., Hu Y., Sasiadek J., 2015. Human friendly autonomous robot using Dempster-Shafer sensor fusion and velocity potential field control. In: *Proceedings of the 11th International Conference on Autonomic and Autonomous Systems (ICAS 2015)*, 24–29 May 2015, Rome, Italy, 92–97.

Neu C., Kirchner E.A., Kim S.K., Tabie M., Linn C., Werth D., 2019. Cognitive work pro- tection -A new approach for occupational safety in human-machine interaction. In: *Information Systems and Neuroscience: Lecture Notes in Information Systems and Organisation*, F. Davis, R. Riedl, J. vom Brocke, P.M. Léger, A. Randolph (Eds.), vol. 29, Springer, Cham, Switzerland. https://link.springer.com/chapter/10.1007/978-3-03 0-01087-4_26#citeas (accessed August 7, 2019).

OECD, 2013. The OECD privacy framework. Organisation for Economic Co-operation and Development (OECD), https://www.oecd.org/internet/ieconomy/privacy-guidelines.ht m (accessed January 2, 2020).

OECD, 2019. Recommendation of the Council on Artificial Intelligence. OECD/LEGAL/0449. Organisation for Economic Co-operation and Development (OECD), https://legalin struments.oecd.org/en/instruments/OECD-LEGAL-0449 (accessed January 2, 2020).

OSHWiki contributors. 2019. Prevention and control strategies, OSHWiki. https://oshwiki .eu/index.php?title=Prevention_and_control_strategies&oldid=241855 (accessed August 1, 2019).

Peniak M., 2019. Real-time PPE monitoring on the edge. Intel-rrk-safety, https://cortexica.gi thub.io/intel-rrk-safety (accessed August 6, 2019).

Perera C., Zaslavsky A., Christen P., Georgakopoulos D., 2013. Context aware computing for the Internet of things: A survey. *IEEE Communications Surveys and Tutorials* 16(1):414–454. https://ieeexplore.ieee.org/document/6512846 (accessed July 30, 2019).

Podgórski D., 2017. Functions of smart PPE in relation to the Hierarchy of Risk Controls. LinkedIn. Posted 11 September 2017. https://www.linkedin.com/pulse/smart-ppe-func tions-vs-classical-hierarchy-risk-daniel-podg%C3%B3rski-1/ (accessed July 20, 2019).

Podgórski D., Majchrzycka K., Dąbrowska A., Okrasa M., Gralewicz G., 2017. Towards a con- ceptual framework of OSH risk management in smart working environments based on smart PPE, ambient intelligence and the Internet of Things technologies. *International Journal of Occupational Safety and Ergonomics: JOSE* 23(1):1–20.

Rakowsky U.K., 2007. Fundamentals of the Dempster-Shafer theory and its applications to system safety and reliability modelling. *Reliability: Theory and Applications* 2(3–4):173–185.

Ranganathan A., Al-Muhtadi J., Campbell R.H., 2004. Reasoning about uncertain contexts in pervasive computing environments. *IEEE Pervasive Computing* 3(2):62–70.

Rashid K.M., Datta S., Behzadan A.H., Raiful H., 2018. Risk-incorporated trajectory prediction to prevent contact collisions on construction sites. *KICEM Journal of Construction Engineering and Project Management* 8(1):10–21. http://www.koreascience.or.kr/artic le/JAKO201812055795549.page (accessed August 6, 2019).

Ridden P., 2017. Dainese testing inflatable protective jacket in the workplace. https://newatla s.com/dainese-d-air-airbag-jacket-enel/51708/ (accessed July 30, 2019).

Sánchez D., Tentori M., Favela J., 2008. Activity recognition for the smart hospital. *IEEE Intelligent Systems* 23(2):50–57. https://ieeexplore.ieee.org/document/4475859 (accessed July 29, 2019).

Sebbak F., Benhammadi F., Chibani A., Amirat Y., Mokhtari A., 2013. Dempster-Shafer theory-based human activity recognition in smart home environments. *Annals of telecommunications – annales des télécommunications* 69(3–4):171–184. https://www.springer professional.de/en/dempster-shafer-theory-based-human-activity-recognition-in-sm art/15613380 (accessed August 6, 2019).

Shafer G., 1976. *A Mathematical Theory of Evidence.* Princeton, NJ: Princeton University Press.

Singla G., Cook D.J., Schmitter-Edgecombe M., 2009. Tracking activities in complex settings using smart environment technologies. *International Journal of Biosciences, Psychiatry and Technology* 1(1):25–35.

Srivastava R.P., Liu L., 2003. Applications of belief functions in business decisions: A review. *Information Systems Frontiers* 5(4):359–378.

Suárez Sánchez A., Riesgo Fernández P., Sánchez Lasheras F., de Cos Juez F.J., García Nieto P.J., 2011. Prediction of work-related accidents according to working conditions using support vector machines. *Applied Mathematics and Computation* 218(7):3539–3552. https://www.sciencedirect.com/science/article/pii/S0096300311011416 (accessed August 5, 2019).

Teizer J., Allread B.S., Fullerton C.E., Hinze J., 2010. Autonomous pro-active real-time construction worker and equipment operator proximity safety alert system. *Automation in Construction* 19(5):630–640. https://www.sciencedirect.com/science/article/pii/S0926 580510000361 (accessed August 6, 2019).

Thanathornwong B., Suebnukarn S., Ouivirach K., 2014. A system for predicting musculoskeletal disorders Among dental students. *International Journal of Occupational Safety and Ergonomics: JOSE* 20(3):463–475.

Tixier A.J.-P., Hallowell M.R., Rajagopalan B., Bowman D., 2016. Application of machine learning to construction injury prediction. *Automation in Construction* 69:102–114. https://www.sciencedirect.com/science/article/pii/S0926580516300966 (accessed August 5, 2019).

Tolstikov A., Hong X., Biswas J., Nugent C., Chen L., Parenteet G., 2011. Comparison of fusion methods based on DST and DBN in human activity recognition. *Journal of Control Theory and Applications* 9(1):18–27.

Tsiakas K., Papakostas M., Theofanidis M., Bell M., Mihalcea R., 2017. An interactive multi-sensing framework for Personalized Human Robot Collaboration and Assistive Training Using Reinforcement Learning. In: *PETRA'17 Proceedings of the 10th International Conference on PErvasive Technologies Related to Assistive Environments*, June 21–23, 2017, Island of Rhodes, Greece, 423–427. https://dl.acm.org/doi/10.1145/3056540.3 076191 (accessed August 6, 2019).

UL (Underwriters Laboratories Inc.), 2017. ANSI/CAN/UL 2900-1:2017, Software cybersecurity for network-connectable products, part 1: General requirements. Underwriters Laboratories Inc., USA.

UNI Global Union, 2017. Top 10 principles for workers' data privacy and protection. Nyon, Switzerland: UNI Global Union. http://www.thefutureworldofwork.org/media/35421/uni_workers_data_protection.pdf (accessed August 8, 2019).

Vinoski J., 2019. IBM Watson announces partnerships to improve Worker safety Through Watson IoT. *Forbes*. https://www.forbes.com/sites/jimvinoski/2019/02/13/ibm-watson-announces-partnerships-to-improve-worker-safety-through-watson-iot/#4a4e82735a8e (accessed July 31, 2019).

Wang D., Dai F., Ning X., 2015. Risk assessment of work-related musculoskeletal disorders in construction: State-of-the-art review. *Journal of Construction Engineering and Management* 141(6), 04015008-1-15. https://ascelibrary.org/doi/abs/10.1061/%28ASCE%29CO.1943-7862.0000979 (accessed August 11, 2019).

Webb A., 2019. The age of smart PPE technology is here. MCR safety. https://www.mcrsafety.com/blog/the-age-of-smart-ppe-technology-is-here (accessed July 31, 2019).

Weber P., Medina-Oliva G., Simon C., Lung B., 2012. Overview on Bayesian networks applications for dependability, risk analysis and maintenance areas. *Engineering Applications of Artificial Intelligence* 25(4):671–682. https://www.sciencedirect.com/science/article/pii/S095219761000117X (accessed August 6, 2019).

Wu H., Zhao J., 2018. An intelligent vision-based approach for helmet identification for work safety. *Computers in Industry* 100:267–277, https://www.sciencedirect.com/science/article/pii/S016636151730461X (accessed August 7, 2019).

Yang J.M., Park J.Y., Im S.Y., Park J.-H., Oh R.-D., 2011. Context-awareness smart safety monitoring system using sensor network. In: *Multimedia, Computer Graphics and Broadcasting, Communications in Computer and Information Science, 263*, 270–277. Berlin Heidelberg: Springer. https://link.springer.com/chapter/10.1007/978-3-642-27186-1_35 (accessed August 6, 2019).

Yang K., Ahn C.R., Vuran M.C., Aria S.S., 2016. Semi-supervised near-miss fall detection for ironworkers with a wearable inertial measurement unit. *Automation in Construction* 68:194–202. https://www.sciencedirect.com/science/article/pii/S0926580516300784 (accessed August 6, 2019).

Ye J., Dobson S.A., McKeever S., 2012. Situation identification techniques in pervasive computing: A review. *Pervasive and Mobile Computing* 8(1):36–66.

Zhang D., Guo M., Zhou J., Kang D., Cao J., 2010. Context reasoning using extended evidence theory in pervasive computing environments. *Future Generation Computer Systems* 26(2):207–216. https://www.sciencedirect.com/science/article/pii/S0167739X09001095 (accessed August 6, 2019).

Zhang J., Yan X., Zhang D., Haugen S., Yang X., 2014. Safety management performance assessment for Maritime Safety Administration (MSA) by using generalized belief rule base methodology. *Safety Science* 63:157–167. https://www.sciencedirect.com/science/article/pii/S0925753513002543 (accessed August 6, 2019).

Index

Page numbers followed by f and t indicate figures and tables, respectively.

For Product Safety Concerns and Information please contact our EU
representative GPSR@taylorandfrancis.com
Taylor & Francis Verlag GmbH, Kaufingerstraße 24, 80331 München, Germany